BRUNEL'S BRITAIN

BRUNEL'S BRITAIN

Derrick Beckett

DAVID & CHARLES
Newton Abbot   London   North Pomfret (Vt)

British Library Cataloguing in Publication Data

Beckett, Derrick
Brunel's Britain.
1. Brunel, Isambard Kingdom
2. Civil engineering
I. Title
624   TA140.B75

ISBN 0–7153–7973–9

Printed in Great Britain
by Biddles Limited Guildford
for David & Charles (Publishers) Limited
Brunel House Newton Abbot Devon

Published in the United States of America
by David & Charles Inc
North Pomfret Vermont 05053 USA

CONTENTS

'I hope you make Brunel your hero'
Prince Philip

PREFACE AND ACKNOWLEDGEMENTS

One hundred and twenty years after Isambard Kingdom Brunel's death in 1859, the Concorde crossed the Atlantic in 2 hours 59 minutes and 36 seconds at an average speed of nearly 1,180mph. It took Brunel's iron steamship, the *Great Britain* over one hundred times as long to complete the same journey, but in relative terms, the time-scales and achievements are similar. There is little doubt that Brunel would have approved of Concorde, but I suspect with a capacity for five hundred not one hundred passengers. This suggests why the commercial world thought him extravagant, but twelve decades after his death, perhaps we need to re-assess commercial opinion.

Currently, the image of engineering is cause for concern at all levels. Prince Philip has suggested that we make Brunel our hero and the Prince of Wales said that many young people might be put off engineering because of its 'rather oily, grimy image'.

In 1977, the Government appointed a Committee of Enquiry into the Engineering Profession which was chaired by Sir Montague Finniston, FRS, and in January 1980, their report 'Engineering our Future' was published at an estimated cost of £0.4 million. The report refers to the misleading national tendency to regard engineering as a subordinate branch of science, the low and declining profitability among British manufacturing companies, and the lack of social standing of engineers in Britain. Innovation and productivity are two key words in the report and are also the key words in the careers of Marc Isambard and Isambard Kingdom Brunel. They achieved immense status in a number of fields of engineering – civil, structural, mechanical and marine – and their ability to translate thought into action underlines the essential difference between science and engineering.

It is my personal philosophy that an appreciation of the role of the engineer in an industrial society can be developed from a perspective of the past, and during the past fifteen years I have presented a course to first year civil engineering students at the University of Surrey which sets out to provide the essential continuity between the past and the present.

History forms an important part of the training of architects, but has not been so common in University Engineering Schools. The course embraces the interaction between architecture and engineering and surveys the works of innovating architects and engineers, including that of the Brunels. It is from this course and the opportunity presented by Dr Les Clark and Ian White to give a talk on the Brunels to the Association of London Graduates and Students of the Institution of Civil Engineers, that the idea for writing this book evolved.

The status of engineering will only improve if the layman and other professions are made aware that it is an art involving imagination, flair and the practical application of scientific knowledge. The 'art' of engineering in its truest sense has rarely been more ably expressed than in the achievements of the Brunels. It is hoped that this book will be of broad interest and make a modest contribution towards wiping clean engineering's oily image. In order to translate thought into action in the form of a book on the Brunels, photographs and illustrations are essential. I am greatly indebted to Frances Gibson Smith, AIIP, and Malcolm Kaye, whose talents and enthusiasm have captured the quality of the work of two remarkable engineers. I am also indebted to Professor Z. S. Makowski, Head of the Department of Civil Engineering, University of Surrey, for constant encouragement, the University's Audio Visual Aids Unit under the direction of Russell Towns, Frank Kietch for the privilege of using his house to write several chapters and to Pauline Wadsworth for typing the text with care and cheerfulness. My wife has endured many moments of desperation and obsession with a man who died over one hundred and twenty years ago. I hope she will forgive me. Finally I would like to acknowledge the opportunity given me by the University of Bristol Library to inspect Brunel's sketch and calculation books, the sensitive guidance of Anthony Lambert in editorial matters and the co-operation of Sir Frederick Snow & Partners in the final stages of this project.

Derrick Beckett
January, 1980

INTRODUCTION

England's architectural heritage from castle to cottage conjures up an arcadian scene which has no room for the indecencies of technology, but the latter half of the eighteenth century was the gestation period for a new architecture which changed the face of England in a mere hundred years. Fortunately, the talents of a number of eminent engineers ensured that much of this new architecture was compatible with that of the past and the natural environment. However, it must be conceded that an appreciation of this new architecture – the architecture of technology – requires rather more effort than that required to enjoy the immediate impact of the thatched cottage, palace, cathedral or castle.

Amongst the group of engineers who pioneered the development of our industrialised society, and in particular civil engineering constructions associated with the rapid growth of transport systems, were William Jessop (1745–1814), Thomas Telford (1757–1834), Marc Isambard Brunel (1769–1849), George Stephenson (1781–1848), Robert Stephenson (1803–59) and Isambard Kingdom Brunel (1806–59). Their works demonstrated admirably the 'art' of engineering in the truest sense, that is, the practical application of scientific knowledge. During the eighteenth century scientific methods were gradually introduced into various fields of engineering and this involved the generation of explanatory laws that could be justified experimentally. In France the use of mathematics in the development of laws to explain the behaviour of structures advanced rapidly, and in 1747 the École des Ponts et Chaussées was founded in Paris for the training of engineers in road and bridge construction. Data relating to the testing of structural materials became available and books on statics and strength of materials were published.

In 1795 the École Polytechnique was opened in Paris (Timoshenko) for the purpose of training engineers in basic sciences as the necessary preparation for entry to schools such as the École des Ponts et Chaussées. The Polytechnique's outstanding teachers, scientists and engineers included Monge, Poisson, Cauchy and Navier. Of French origin, the

Brunels were naturally acquainted with current scientific progress and in 1822 Isambard Kingdom Brunel attempted unsuccessfully to enter the École Polytechnique. In later life he made the following comment on technical education:

I must strongly caution you against studying practical mechanics among French authors – take them for abstract science and study their statics, dynamics, geometry etc etc, to your heart's content, but never even read any of their works on mechanics any more than you would search their modern authors for religious principles. A few hours spent in a blacksmiths and wheelwrights shop will teach you more practical mechanics – read English books for practice – there is little enough to be learnt in them but you will not have to unlearn that little.

(I. K. B. 2 December 1848)

The Brunels were primarily professional engineers, and although France was ahead of England in theoretical studies, this was over-shadowed by England's superiority in industrial and commercial techniques and most important of all, in mechanical inventiveness.

It has been argued that the Middle Ages was one of the great inventive eras of mankind (Gimpel) and that it should be known as the first industrial revolution in Europe. A parallel is drawn between the watermill and the steam engine, but in the author's view, the first real outburst of mechanical inventiveness occurred in England during the eighteenth century. The factors contributing to this are numerous, (Hartwell) and, during a period of a hundred years, Britain was the place where a dozen or more basic inventions were developed and introduced into industry. This led to fundamental changes in methods of transport and in the manufacturing industries. It was fortuitous that when Marc Brunel arrived in England at the end of the eighteenth century, some of these fundamental changes had taken place or were in gestation. Thus the conditions were favourable to allowing him to develop his ingenuity in a way which had a significant influence on nineteenth-century manufacturing techniques. This was followed by his son's immense contribution to transport by rail and ship.

Perhaps the most significant of the eighteenth-century innovations was the development of steam power. An acute timber famine in England led to efforts to replace it by coal in the iron industry, and by 1750 coal-mining had become a well-established industry in England and western Europe. The Newcomen stationary condensing steam engine helped to solve the problem of mine drainage which encouraged increased production and further expansion of the iron industry. Subsequent advances in the use of steam as a source of power and locomotion were influenced by James Watt (Singer *et al*), Richard Trevithick and George Stephenson. This gave Isambard Kingdom Brunel the basic technology which, when combined with his sound education in theoretical matters, provided a unique opportunity to make possibly the greatest contribution to nineteenth-century industrial architecture during an independent professional life of thirty years between 1829 and 1859.

Brunel's most famous structures are the elegant railway terminus at Paddington, which on 1 March 1979 celebrated its 125th anniversary, the majestic Royal Albert Bridge at Saltash, spanning the River Tamar to form the rail link between Devon and Cornwall, and the spectacular suspension bridge over the Avon Gorge at Bristol, completed five years after his death. However, many of Brunel's other works, still functioning efficiently today, are much less widely known.

The use of British Rail's High Speed Trains (HST) on the straight, well-graded 118 mile route from Paddington to Bristol Temple Meads — Brunel's billiard table — exemplifies his conviction that 'speed within reasonable limits is a material ingredient in perfection in travelling'. Passengers sitting in air-conditioned comfort and travelling at 125mph will obtain no more than a fleeting glance of his numerous engineering works. The most outstanding of these are the Wharncliffe Viaduct at Hanwell, Maidenhead Bridge, Sonning Cutting, the railway works and village at Swindon, Box Tunnel, the bridges and viaduct at Bath and the train shed at Bristol Temple Meads.

It is the primary aim of this book to chronicle the engineering works of Isambard Kingdom Brunel as they exist today and to examine their technical and architectural merit, which will hopefully encourage the reader to inspect what is left of the 'finest work in England', engineered with an enthusiasm, imagination and flair the like of which has rarely, if ever, been equalled. In a tribute to Brunel, Daniel Gooch, his locomotive engineer, wrote that 'the commercial world thought him extravagant, but although he was so, great things are not done by those who count the cost of every thought or act'.

It is inevitable that comparisons are made between the Stephensons and the Brunels, and in his *Lives of the Engineers* Samuel Smiles states that the Stephensons were inventive, practical and sagacious and the Brunels, ingenious, imaginative and daring. Isambard Kingdom Brunel was determined to produce the best constructions that imagination could devise, whereas the main objective of the Stephensons was to show profitable results from their engineering works. However, to demonstrate that his Great Western Railway had been economically built, Brunel made the observation that 'we look sharper after money than you suppose though perhaps we don't talk about it so much down here as you do up in the north'. (I. K. B. 20 July 1842)

Smiles concluded that, measured by practical and profitable results, the Stephensons were unquestionably the safer men to follow, and it could be reasoned that Robert Stephenson's Britannia Tubular Bridge, crossing the Menai Straight and completed in 1850, was technically the most significant of nineteenth-century bridge structures in terms of its influence on recent developments. Its structural form was the product of combined theoretical and experimental work on stiffened iron plates leading to the first use of a box girder for a major civil engineering project. The box girders were able to support their own weight of 1,600 tons for the central 460ft (142m) spans plus that of the trains without the assistance of suspension chains as envisaged prior to experimental investigations. In contrast, Brunel's design for the Royal Albert Bridge, Saltash, of similar span to the Britannia Bridge, was an ingenious combination of a tubular arch and chain suspension system which resulted in a total cost which compared very favourably with that for the Britannia Bridge. However, there are other factors to be considered and the Royal Albert Bridge is described elsewhere.

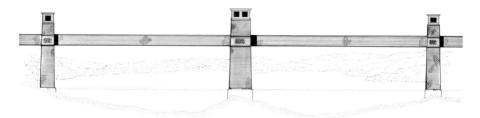

Fig 1 Britannia Tubular Bridge, designed by Robert Stephenson and opened in 1850

Not wishing to detract in any way from the immense contribution of the Stephensons to the development of civil and mechanical engineering in the nineteenth century, the combination of Brunel's charisma, productivity, flair and attention to architectural detail is a subject of great fascination. Currently, interest in the Brunels is approaching a state of mania as demonstrated by exhibitions, restoration projects and so on. The celebration of the 125th anniversary of the opening of Paddington Station included the issue of a souvenir booklet and the staging of an exhibition and photographic display. The author was recently involved in devising an exhibition in conjunction with the Audio

Isambard Kingdom Brunel standing before the launching chains of the *Great Eastern* in 1858 (*National Portrait Gallery*)

Visual Aids Unit of the University of Surrey which is a photographic illustration of the railway from Paddington to Penzance. The Borough of Thamesdown is restoring the railway village at Swindon and a voluntary association is restoring Marc Brunel's engine house, chimney and shaft at Rotherhithe. A staggering £7,500 was paid by the Victoria and Albert Museum for the classic photograph of Isambard Kingdom Brunel which shows him standing before the launching chains of the PSS *Great Eastern*, one of the first iron steamships built for regular transoceanic crossings. It dates from around 1858 and was one of a set of prints from negatives commissioned by the *Illustrated Times*.

In order to set the background to a more detailed explanation of why this interest in the Brunels has evolved, a brief survey of their lives and achievements is desirable.

CHRONICLE OF EVENTS

Date	
1747	École des Ponts et Chaussées founded in Paris for training engineerings in construction work on highways, tunnels and bridges
25 April 1769	Marc Isambard Brunel born in Hacqueville, Normandy
1795	École Polytechnique opened in Paris for training engineers in basic sciences
13 March 1799	Marc Brunel arrives in England
9 April 1806	Isambard Kingdom Brunel born in Portsmouth
23 May 1822	First rail of Stockton & Darlington Railway laid (Stephensons)
21 August 1822	I. K. Brunel completes education at Lycée Henri-Quatre in Paris returning to England after an abortive attempt to enter École Polytechnique
1823	I. K. Brunel regularly employed in his father's office and engaged on work connected with Thames Tunnel
27 November 1824	Railway proposed between Bristol and London and adopted at a meeting in a London tavern
27 September 1825	Opening of Stockton & Darlington Railway (Stephensons)
December 1825	Thames Tunnel works under direction of I. K. Brunel
28 January 1828	Second major accident put stop to work on the Thames Tunnel for 7 years
10 June 1830	I. K. Brunel elected Fellow of the Royal Society
15 September 1830	Opening of Liverpool & Manchester Railway (Stephensons)

31 March 1831	I. K. Brunel appointed Engineer to Clifton Bridge
7 March 1833	I. K. Brunel appointed Engineer to Great Western Railway (GWR)
August 1835	First proposal to build a steamship for Bristol–New York run at meeting of GWR, Radleys Hotel, Blackfriars, London
29 October 1835	I. K. Brunel's proposal for a 7ft broad gauge rail adopted
5 July 1836	I. K. Brunel married Mary Horsley. Three children – Isambard, Henry Marc and Florence Mary
27 August 1836	Foundation stone for Leigh abutment of Clifton Bridge laid
19 July 1837	Launch of the steamship (PS) *Great Western* (wooden hull)
28 November 1837	Delivery of *North Star* locomotive to GWR (2–2–2, 7ft driving wheel)
4 June 1838	Opening of section 1 of GWR, Paddington to Maidenhead (old stations)
1 July 1839	Opening of section 2 of GWR, Maidenhead to Twyford
19 July 1839	Keel plates laid for SS *Great Britain* (metal hull)
30 March 1840	Opening of section 3 of GWR, Twyford to Reading
1 June 1840	Opening of section 4 of GWR, Reading to Steventon
20 July 1840	Opening of section 5 of GWR, Steventon to Faringdon Road
31 August 1840	Opening of section 6 of GWR, Bristol (old terminus) to Bath
17 December 1840	Opening of section 7 of GWR, Faringdon Road to Hay Lane
31 May 1841	Opening of section 8 of GWR, Hay Lane to Chippenham
30 June 1841	Completion of GWR, London to Bristol
24 December 1841	Accident at Sonning Cutting, 8 killed, 17 injured
1 July 1842	Opening of Bristol to Taunton section of Bristol & Exeter Railway (B&ER)

1 February 1843	Swindon locomotive works opened
25 March 1843	Thames Tunnel opened to public 18 years after commencement of work
1 May 1843	Opening of Taunton to Beambridge section of B&ER
19 July 1843	Launch of the SS *Great Britain*
1 May 1844	Opening of Beambridge to Exeter section of B&ER
9 August 1844	Gladstone's Railway Act – compulsory to provide accommodation for at least one train each day on weekdays at 1d per mile (3rd class)
18 August 1846	Act for regulating gauge of railways received the Royal Assent
23 September 1846	*Great Britain* ran aground Dundrum Bay, Co Down
30 December 1846	South Devon Railway extended from Teignmouth to Newton Abbot
August 1847	SS *Great Britain* refloated
13 September 1847	Atmospheric trains began to carry passengers between Exeter and Teignmouth, highest speed recorded 68mph with 28-ton train
10 January 1848	Atmospheric working extended to Newton Abbot
5 May 1848	The South Devon Railway opened as far as Laira Green within 3 miles of Plymouth
2 April 1849	The South Devon Railway completed to Plymouth terminus between Union Street and Millbay Road
8 October 1849	Windsor branch of GWR opened
12 December 1849	Marc Brunel died in 81st year
11 March 1852	West Cornwall Railway opened from Penzance to Hayle
25 August 1852	West Cornwall Railway opened from Redruth to Truro
16 January 1854	Paddington New Station opened (departure side)
1 May 1854	First plates of PSS *Great Eastern* laid (double-skin metal hull)
29 May 1854	Paddington New Station opened (arrival side)
1 February 1855	World's first postal train London to Bristol

16 February 1855	Work commenced on Renkioi hospital, Dardanelles
7 May 1855	First vessel carrying prefabricated units arrives Dardanelles
21 May 1855	Erection of buildings at Renkioi commenced
12 July 1855	Hospital at Renkioi ready for 300 sick
1857	PS *Great Western* broken up at Vauxhall
3 November 1857	First launch of PSS *Great Eastern* attempted
31 January 1858	PSS *Great Eastern* launched
11 April 1859	Royal Albert Bridge, Saltash completed
6 September 1859	PSS *Great Eastern* advertised to sail
15 September 1859	Isambard Kingdom Brunel dies
20 September 1859	I. K. Brunel buried in Kensal Green Cemetery
June 1860	PSS *Great Eastern* commences first Atlantic run
8 December 1864	Opening ceremony for Clifton Suspension Bridge
1 March 1867	First through passenger service from London to Penzance
20 May 1892	Last broad gauge trains ran on GWR
10 April 1979	Western Region of British Rail establishes world record for a start to stop average speed – average speed 111.6mph on 94-mile run from Paddington to Chippenham

1
THE BRUNELS

Marc Isambard Brunel (1769–1849)

I had liked the son, but at our very first meeting I could not help feeling that his father far excelled him in unworldliness, genius and taste (Charles Macfarlane)

Lady Gladwyn, the great grand-daughter of Isambard Kingdom Brunel and grand-daughter of Florence Mary Brunel, suggested in a paper presented to a joint meeting of the Institution of Civil Engineers and Société des Ingénieurs Civils de France, that to differentiate between the father and son might be done by referring to Marc Isambard and Isambard Kingdom. The simplified Brunel family tree will hopefully clarify any misunderstanding (Fig 2).

Marc Isambard Brunel, the son of Jean Charles Brunel and Marie Victoire Lefèvre was born in the hamlet of Hacqueville in northern France on 25 April 1769. Jean Charles Brunel was a prosperous farmer and a recent photograph of 'La Ferme Brunel' (Clements) indicates that it was partially rebuilt in the nineteenth century. Marie Victoire Lefèvre, the second of the four wives of Jean Charles Brunel, was the niece of the Abbé Lefèvre, the godfather to Marc Isambard. The name Isambard was apparently reserved for members of the family destined to enter the church, but for the Abbé's godson, this was not to be.

At an early age, Marc Isambard displayed a talent for drawing, mathematics and mechanics. To his father's displeasure, his eleven-year-old son declared that he wished to become an engineer. This prompted his father to send him to a seminary in Rouen, but fortunately the Superior recognised that Marc Isambard's talents could best be developed elsewhere, and arrangements were made for him to lodge with an elder cousin and be tutored for eventual entry to the navy as an officer cadet. One of his tutors was Gaspard Monge (1746–1818) who later played a leading part in the setting up of the École Polytechnique. Monge was a gifted mathematician and teacher and thus Marc Isambard had the

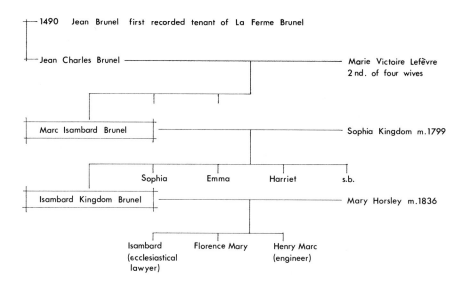

```
┬─1490   Jean Brunel   first recorded tenant of La Ferme Brunel
│
│
├─Jean Charles Brunel ──────────────────────────────────────── Marie Victoire Lefèvre
│                                                                2 nd. of four wives
│
│
│        Marc Isambard Brunel ─────────────────────────────── Sophia Kingdom m.1799
│
│                    Sophia      Emma        Harriet       s.b.
│        Isambard Kingdom Brunel ───────────────────────────── Mary Horsley m.1836
│
│                Isambard        Florence Mary    Henry Marc
│                (ecclesiastical                  (engineer)
│                lawyer)
```

Fig 2 Simplified Brunel family tree

benefit of the best mathematical training available in Europe at that time. In 1786, the seventeen-year-old cadet was able to embark on a naval career which lasted six years. He was paid off in 1792 and in the following eighteen months his royalist sympathies made life hazardous, but during this time he met a young English girl, Sophia Kingdom, his future wife. In July 1793 he was forced by the events of the French Revolution to flee to America, leaving Sophia in Rouen.

It was in New York that Marc Isambard was first able to apply his engineering talents in a practical way, and in 1796 became the city's Chief Engineer. Two years later, a dinner party conversation involved the discussion of various aspects of warship design, including the Royal Navy's requirement for 100,000 ship's pulley blocks a year. This was a challenge to Marc Isambard's mechanical inventiveness and subsequently he resigned from his position and armed with a letter of introduction to the Navy Minister, sailed for England on 7 February 1799. It is of interest to speculate whether on the tedious journey via Falmouth and Plymouth to London, his fertile imagination dwelt on more efficient means of motive power than the sail or horse. Within his lifetime this became a reality – his son completed the railway line from Paddington to Plymouth and built a screw-driven iron steamship, the SS *Great Britain*, which crossed the Atlantic in less than fourteen days. On reaching London, Marc Isambard was able to join Sophia Kingdom and their marriage took place in November 1799.

The following twenty-five years was a period of intense activity in which sixteen patent specifications were filed including machines for

Marc Isambard Brunel by S. Drummond (*National Portrait Gallery*)

writing and drawing, ship's blocks, machines for sawing timber and cutting veneers, shoes and boots, sawmills, a tunnelling machine, marine steam engines and gas engines. All of these contributed to the development of manufacturing techniques during the nineteenth century.

In 1801, Samuel Bentham, who was at that time Inspector General of Naval Works, was approached by Marc Isambard with a scheme for manufacturing ships's blocks. The ship's block (Fig 3) consisted of a pulley turning on a pin in a shell. Working models of machines for the production of the components were manufactured by Henry Maudslay (1771–1831), one of the outstanding mechanics of the early nineteenth century. Bentham ordered production machines and Marc Isambard moved to Portsea, a short distance from the dockyard. This pioneering enterprise in mass production was soon followed by further inventiveness, including the mechanical production of shoes and boots. A factory was set up at Battersea and the production level eventually reached 400 pairs of shoes and boots per day. In the period 1811–26, the Brunel's family home was at Lindsey House in Cheyne Walk, Chelsea, a short distance from the factory at Battersea.

Fig 3 The components of a ship's block

During the first decade of the nineteenth century, shipbuilding requirements and the depletion of English forests necessitated importing increasing quantities of timber, and the problem of handling and converting it gave Marc Isambard another opportunity to exercise his inventiveness. In 1812 he was requested by the Admiralty to prepare plans for a log-handling and sawmilling plant to be constructed at Chatham Dockyard. The plant consisted of a pitched-roof sawing hall with two open sides, a boiler house and a waterway (part cutting and part tunnel), connecting the sawmill via a lagoon to the River Medway. Logs were floated via the waterway to a shaft in front of the sawmill. They

The sawmill, Chatham. The once-open sides between the columns have been filled in to form a store (*Frances Gibson Smith*)

were then raised to the surface and mechanically handled into position in front of the saw frames. On conversion of the logs to planks, they were transferred by means of a rope-worked incline railway to an open seasoning area and then a covered store. There is evidence that the rope railway was of 7ft gauge (Clements) and thus, may have been the origin of Isambard Kingdom's broad gauge railway from Paddington to Bristol (see Chapter 3). Fortunately, the sawmill, boiler house and chimney are listed buildings and form part of an impressive collection of Georgian dockyard structures at HM Naval Base, Chatham. The plant demonstrates admirably the catholic nature of Marc Isambard's engineering talents in an operation which combined hydraulics, mechanical handling, steam power and structural design.

In spite of further inventions and schemes for the design of bridges in England, France and Russia, Marc Isambard's financial affairs were never very sound, and in 1821 a series of events led to his consignment to prison, together with his wife Sophia. However, in recognition of his

The massive timber roof trusses at the sawmill in Chatham, showing joint details (*Frances Gibson Smith*)

extraordinary talents and benefit to the country, he was released after a few months. Over three years before, he had filed a patent specification for forming tunnels or drifts underground which, it is said, was inspired by observations of the action of ship worms (*Teredo navalis*) while working at Chatham Dockyard. This device led to the successful tunnelling under the Thames between Rotherhithe and Wapping. He was appointed Engineer to the Thames Tunnel Company in 1824. The ceremonial stone was laid in the Rotherhithe shaft on 2 March 1825, but the tunnel was not opened to foot passengers until 25 March 1843. This monumental achievement in civil engineering (described in Chapter 10) occupied over eighteen years of Marc Isambard's life and five months before its completion, he suffered his first stroke. By this time his son Isambard Kingdom, who worked with him on the early stages of the construction of the tunnel, had achieved even greater fame in railway and ship construction. Supported by Isambard Kingdom during the final years, Marc Isambard died on 12 December 1849, aged eighty four years after a second stroke which had left him partially paralysed. Charles Macfarlane stated that Marc Isambard 'lived as if there were no rogues in this nether world' (Clements) but his son did not suffer from that beautiful illusion.

Isambard Kingdom Brunel (1806–59)

I am always building castles in the air, what time I waste (I. K. B.)

It is a rare event for castles in the air to materialise and it is fortunate for us that many of Isambard Kingdom Brunel's 'castles' are with us today and represent some of the finest examples of Victorian engineering and architectural achievements. Isambard Kingdom was born on 9 April 1806 at Portsea at the time his father was involved in the development of block-making machinery for the Navy. He inherited his father's talent for drawing and mathematics and in the early formative years benefited immensely from the tuition Marc Isambard bestowed with such benevolence. Some time was spent at a boarding school in Hove, and in 1820 his father sent him to the College of Caen in Normandy. This was followed by a period at the College Henri-Quatre in Paris, famous for its mathematics teachers, and finally by a period of apprenticeship under the talented instrument maker, Louis Breguet.

Naturally, Isambard Kingdom was acquainted with his father's work and from 1823 became involved in the Thames Tunnel project which was shortly to dominate their lives. It was during the early stages of its construction that he kept a private diary which has revealed aspects of his complex nature – bouts of dispiritedness and elation and of course his *châteaux en Espagne*. However, one can sense a latent enthusiasm and energy which was soon to give birth to an exponential growth of productivity which has rarely been equalled. The opportunity presented itself at Bristol in 1829. As a result of the second major flooding of the Thames Tunnel, Isambard Kingdom sustained a severe injury and the

Isambard Kingdom Brunel in 1857 by J. C. Horsley (*National Portrait Gallery*)

latter part of his convalescence was spent at Clifton on the outskirts of Bristol. A proposal for a bridge to span the Avon Gorge came to his notice and there was an invitation for engineers to submit designs. It is difficult to imagine a more romantic setting for a bridge, ideal for the twenty-four-year-old engineer to exercise his imagination, and for the first time as an independent engineer. In all, twenty-two designs were submitted, four by Isambard Kingdom, and Thomas Telford was called in to adjudicate. He rejected all the designs and was then asked to propose his own. Telford's design, with its colossal Gothic-style piers, clashed disastrously with the superb landscape of the Avon Gorge, but fortunately public opinion was against it and it was decided to hold a second competition. Twelve designs were selected for final assessment, excluding Telford's and including Brunel's. Isambard Kingdom's design satisfied the assessors' requirements that the stress in the suspension chains should not exceed 5.5 tons per square inch and on 26 March 1831 he was formally appointed to design and construct the bridge. Clifton Bridge was referred to by Isambard Kingdom as 'my first child, my darling', and he was involved in its design over a period of thirty years until his death in 1859, but due to financial difficulties the two piers had not been completed. Further modifications were made to the final design (see Chapter 6) and the bridge was completed in 1864.

In 1833 he was appointed engineer to the Great Western Railway between Paddington and Bristol, and an impression of the work load imposed on Isambard Kingdom and his small band of assistants can be gained from this extract from *The Life of Isambard Kingdom Brunel* by his son, I. Brunel, published in 1870:

> His own duty of superintendence severely taxed his great powers of work. He spent several weeks travelling from place to place by night and riding about the country by day, directing his assistants and endeavouring, very frequently without success, to conciliate the landowners on whose property he proposed to trespass. His diary of this date shows that when he halted at an inn for the night but little of his time was spent in rest, and that often he sat up writing letters and reports until it was almost time for his horse to come round to take him on a day's work.
>
> 'Between ourselves,' he wrote to Hammond, his assistant, 'it is harder work than I like. I am rarely much under twenty hours a day at it.'

Much of the twenty hours was spent scurrying between various sites in his travelling carriage (britzka) which was designed to carry plans, engineering instruments and of course, a box of fifty cigars. He required a high standard of gentlemanly conduct from his assistants and had no time for idleness: 'I have the impression – if it is wrong correct me – that

The designs by Brunel and Telford for the Clifton Bridge at Bristol

he is one of those who gets up late, to go to their work at gentlemanly hours – and from whom it is difficult to get any real work.'

By December 1835, his professional jobs involved a capital of over £5.5 million, but in spite of this his diary revealed the continuing melancholy: 'Everything has prospered, everything at this moment is sunshine. I don't like it – it can't last – bad weather must surely come. Let me see the storm in time to gather in my sails. Mrs B – I foresee one thing – this time 12 months I shall be a married man. How will that be? Will it make me happier?'

On 5 July 1836 Isambard Kingdom Brunel married Mary Horsley and

An artist's impression of Brunel's design for the Clifton Bridge c. 1831. Note the inward facing sphinxes, which on other views, are shown facing outwards. This impression should be compared with the structure as built in 1864, pp 104–5.

there were three children, Isambard, Henry Marc and Florence Mary. Isambard Brunel was a partial cripple and entered the legal profession, whereas Henry became a competent engineer and built the Connell Ferry Bridge in Argyll. He also assisted with the Tower Bridge, completed in 1894. However, it was Isambard Brunel who wrote his father's biography which was published in 1870. The Isambard Kingdom Brunels maintained an elegant lifestyle in contrast to the more homely elder Brunels. In spite of his punishing work load, Isambard Kingdom was a devoted father and his charm and wit permeated the Duke Street household.

At the time Isambard Kingdom and Mary Horsley were married, construction of the Great Western Railway had commenced but the numerous civil engineering works on the nine sections of the line between Paddington and Bristol were not completed until June 1841. In the middle of this work, Isambard Kingdom proposed to build a steamship for the Bristol-New York run and in July 1837 the wooden-hulled PS *Great Western* was launched. Two years later the keel plates for the iron-hulled SS *Great Britain* were laid. By 1854 he had been involved in the construction of over 1,000 miles of railway, and it was in this year that construction of the PSS *Great Eastern* commenced. She remained the

Connell Ferry Bridge, Argyll, built by Henry Brunel

world's largest ship until the end of the nineteenth century, and the problems involved in its financing, design and construction contributed to his death in 1859.

Considering his railway and maritime commitments, it is difficult to envisage that Isambard Kingdom's work capacity could be sufficient to cope with further engineering problems, yet they included docks and harbours, gunnery experiments, a floating gun carriage, projects overseas and a prefabricated hospital. Perhaps the most intriguing of these is the work on the prefabricated units which were shipped from England and erected at Renkioi on the Dardanelles. Due to the energy and perseverance of Florence Nightingale, the British Government was forced to pay heed to the condition of the troops in the Crimea and Isambard Kingdom was asked to prepare a design for a prefabricated hospital structure. The brief was for a set of buildings of cheap and portable construction, to accommodate up to 1,500 beds capable of being erected on, within reason, any plot of land. Brunel's solution consisted of a covered way extending for about one third of a mile on each side of which were ward units 100ft × 40ft which are shown in longitudinal and transverse section in Fig 4. The structural framework was in timber and innovations included a polished tin roof to give high solar reflection, tarred wooden sewers, mechanical ventilation, provision for the addition of an internal lining of boarding with cavity insulation and an internal limewash finish with a colour tint to reduce glare. Erection of the buildings commenced on 21 May 1855 and the hospital was ready for 300 sick on 12 July 1855.

Renkioi Hospital and Joseph Paxton's Crystal Palace are arguably two of the most significant nineteenth-century building structures in that they incorporated, on a large scale, the essential feature of industrialised building, that is the transfer of the focal point of the building process from the site to the factory. Industrialised building techniques form an essential part of the twentieth-century building philosophy, and prefabricated components are now used extensively for domestic, industrial and bridge structures. Architects and engineers involved in the design of buildings to be constructed in hazardous environments would benefit from studying Isambard Kingdom's description of the Renkioi buildings which is included in the biography by his son.

Five months after the completion of the Royal Albert Bridge at Saltash on 11 April 1859 and seven days after sending the PSS *Great Eastern* to sea, the fifty-three-year-old engineer, prematurely aged, received the final blow – news of an explosion which severely damaged the great steamship, although not to a significant extent structurally. Isambard Kingdom died on 15 September 1859 and was buried in Kensal Green Cemetery.

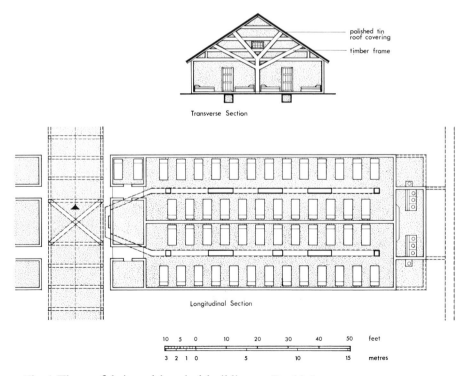

Fig 4 The prefabricated hospital building at Renkioi

For thirty years he devoted all his energy to the art and practice of professional engineering, making full use of the technology currently available and in particular, experimental work. He had little time for state intervention in the form of design rules and was contemptuous of state honours. In contrast to his father, he did not file any patent specifications and spoke at length on the disadvantages of patents with regard to the progress of technology.

2
THE ERA OF RAILWAYS

In 1815 the Duke of Wellington returned from the Battle of Waterloo to an England in which the majority of workers were still in occupations connected with agriculture, but the influence of industrial development was beginning to take effect. Inventions in the field of mechanical engineering such as the spinning jenny, the double acting steam engine, the steam locomotive, and numerous machine tools developed by Bramah, Bentham, Maudslay and Marc Isambard Brunel, which took place in the latter part of the eighteenth and beginning of the nineteenth centuries, led to the gradual replacement of hand methods of production by the machine. This resulted in rapid expansion of the textile, iron, coal and engineering industries. Associated with this industrial development was a dramatic increase in urban populations and the subsequent need for improved means of transport. Although by 1830, a network of canals had virtually been completed and there was a vast improvement in the standard of construction of roads, this was not enough to meet the insatiable demands of industry.

Due to conflicting interests, proposals for the construction of railways frequently met with fierce resistance, in particular from canal companies, turnpike trusts and coaching interests that ran the roads and, of course, from all enemies of progress of any kind. Colonel Sibthorpe, MP for Lincoln, declared, 'I would rather meet a highwayman, or see a burglar on my premises than an engineer.' However, the demands of industry and commerce were such that the battle was won by the railways, and their construction stimulated employment in other industries as well as their requirement for gangs of navvies, drivers, firemen, supervisory staff and engineers. By 1845 it was estimated that upwards of 200,000 men were working on over 3,000 miles of line and the engineering of the works was on a scale which dwarfed the monumental constructions of the past.

The Pre-Locomotive Era

The principle of a railway is to form a track laid with rails to guide vehicles travelling along it, which at the same time provides a smooth contact surface between the wheels and rails to reduce friction. This means that less effort is required to move a load, which for modern British freight trains can exceed several thousand tons.

The origin of railways in which wagons are pulled by horses along tracks is a matter for conjecture, but wagons with wooden flanged wheels running along wooden tracks were in use in the sixteenth century. Subsequently, strips of iron were fixed to the top of the wooden rails to prevent wear and later various shapes of iron rail were designed to replace wooden rails including T, L & I sections. Basically there were two types of railways in use at the end of the eighteenth century – the edge rail of T or I section, which was used in conjunction with a flanged iron wheel, and the L-section plate rail with the vertical flange on the inner edge, not on the wheel (Fig 5). At the beginning of the nineteenth century, the plate rail was considered to be advantageous because wagons could easily be moved off the rails on to normal roads, but there were practical and technical disadvantages. The distance between the rails was governed by the width required for a horse to walk within them

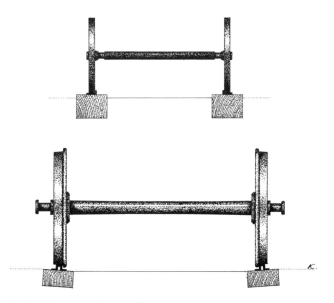

Fig 5 Edge and plate rails

while pulling a wagon. This was in the order of 4ft. The term 'gauge' is used to define this distance and the 'standard gauge' of 4ft $8\frac{1}{2}$in was first established on the Willington Colliery wagonway system near Newcastle upon Tyne. George Stephenson was involved with part of this system which embraced Killingworth Colliery. Early developments in railways are generally associated with the north-east of England but the world's first public goods railway to be sanctioned by Parliament, the Surrey Iron Railway, was opened from Wandsworth to Croydon on 26 July 1803. It was designed by William Jessop, of canal fame (Hadfield and Skempton), and the track consisted of 3ft span L-section plate rails spanning between stone blocks (see Technical Appendix). The gauge was 4ft 2in and the wagons had a gross weight in the order of 3 tons.

Nine years after the opening of the Surrey Iron Railway, the first commercially used steam locomotives, designed by John Blenkinsop and built by Matthew Murray, ran on the Middleton Colliery Railway, Leeds. The locomotive was propelled by a toothed wheel engaging on a rack on one of the rails. This was followed by the inventive genius of George Stephenson who, in 1815, built the Killingworth locomotive which was reported to have pulled a total load of 50 tons on level ground at 6mph. A single horse is capable of pulling a load of about 15 tons at a somewhat lower speed. A detailed description of the development of the locomotive as a reliable form of motive power is given in Chapter 9.

The first public railway to use steam from the beginning was the Stockton & Darlington Railway, constructed by George Stephenson and opened on 27 September 1825. Just under five years later, the Duke of Wellington (then Prime Minister) attended the opening of the Liverpool & Manchester Railway (15 September 1830), the first public railway to be operated almost entirely by locomotives and the first to operate passenger trains to a timetable. From 1830 onwards, the steam locomotive, described by the Rev W. Awdry as 'the most human of man-made objects', rapidly replaced the horse as a means of motive power. Thus Isambard Kingdom Brunel, at the commencement of his involvement with the construction of railways, could draw upon over three decades of continuous development of the steam locomotive but, as we shall see later, his first encounter with locomotive design proved to be a disaster.

The Great Western Railway (1824–35)

On 27 December 1824, some nine months prior to the opening of the Stockton & Darlington Railway, a proposal for the formation of the London & Bristol Rail-Road Company was formally adopted at a meeting held in the London Tavern, and at this time the eighteen-year-old Isambard Kingdom Brunel was employed in his father's office, no doubt heavily involved with work on the Thames Tunnel. At a further meeting it was resolved to apply for an Act of Parliament, but the application was not made and the project lapsed for a number of years. However, in 1832, the project was revived by a group of prominent business men who were no doubt spurred on by previous events in the North, in particular, the opening of the world's first inter-city line between Liverpool and Manchester. Funds were provided for a preliminary survey and estimate, and a committee was set up for the selection of an engineer. From several candidates, Isambard Kingdom Brunel was appointed and immediately embarked on detailed surveys, a number of routes being investigated.

Obtaining an act of incorporation for a railway is a tedious process as a plan and section of each part of the land through which the intended line is to pass has to be drawn up and made available for inspection by the interested parties. Between 1825 and 1837, ninety-three railway acts of various types went through Parliament but possibly more than any other, the Great Western Railway bill was subjected to the most detailed examinations. The bill was brought into the House of Commons in 1834 and the committee investigation lasted fifty-seven days. However, Isambard Kingdom Brunel's charisma survived an eleven-day cross-examination and an engineer who witnessed the event remarked, 'I do not remember having enjoyed so great an intellectual treat as listening to Mr Brunel's examination.' The bill was eventually passed by the Commons but rejected by the Lords on the second reading. This was largely due to the fact that the bill was for only part of the line between London and Bristol, and subsequently a new bill for the complete line was investigated by the Commons committee in early 1835. It withstood further fierce opposition and reached the Lords where it was introduced by Lord Wharncliffe. Following forty days of committee examination, the bill received Royal Assent on 31 August 1835, and the Great Western Railway Company was empowered to construct a railway between London and Bristol with Isambard Kingdom Brunel as Engineer. Brunel responded immediately and on 3 September 1835 he embarked on setting out a line between Bristol, Bath, Reading and London. Until 1843 the affairs of the Great Western Railway were managed by separate London and Bristol committees. He states in a letter that 'we shall have our flags flying over the Brent valley tomorrow. I should not wish that Bristol should fancy itself left behind. I shall be down on Tuesday or Wednesday.' The 'finest work in England' was thus underway.

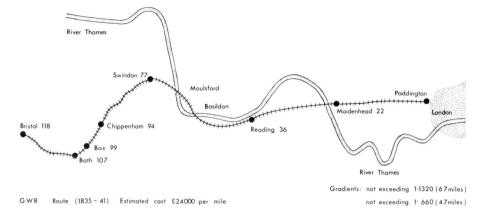

Fig 6 The route chosen for the Great Western Railway

The route finally chosen for the Great Western Railway commenced in a field lying between the Paddington Canal and the turnpike road leading from London to Harrow, passing over the Brent Valley via the Wharncliffe Viaduct and then through Slough and on to Maidenhead where it crosses the Thames at a distance of 22 miles from London (Fig 6). It continues in a westerly direction to Reading, 36 miles from London and then takes a north westerly direction, crossing the Thames at Basildon and Moulsford. At Moulsford it departs from the northerly direction of the Thames and continues west on to Steventon and then Swindon, 77 miles from London. Diverting in a south westerly direction through Chippenham and Bath, it crosses the Avon and then on to Bristol where it once terminated in a field called Temple Mead, 117 miles from London.

The works proceeded simultaneously from London and Bristol and the line was constructed in nine sections, the order of completion being listed in Table 1. (Macdermot)

	SECTION	DATE OPENED
1	Paddington to Maidenhead	4 June 1838
2	Maidenhead to Twyford	1 July 1839
3	Twyford to Reading	30 March 1840
4	Reading to Steventon	1 June 1840
5	Steventon to Challow	20 July 1840
6	Bristol to Bath	31 August 1840
7	Challow to Hay Lane	17 December 1840
8	Hay Lane to Chippenham	31 May 1841
9	Chippenham to Bath	30 June 1841

Table 1 Opening dates for the nine sections of the line

CONSTRUCTION

Railway engineers were faced with construction problems of a very different order of magnitude than those associated with turnpike roads (a turnpike is a gate which stops vehicles passing before a toll is paid). Turnpike roads could follow the contours of the land, and gradients of 1 in 12 or steeper were not uncommon – even today, a gradient of 1 in 25 is an acceptable design standard for a rural motorway in hilly country. The gradient is defined as the ratio of the distance the road or rail rises in a specified horizontal length (Fig 7). Experiments indicated that on a smooth level road surface a force of about 60lb would pull a wagon weighing one ton whereas on a level railway, this reduced to 8lb. If we now consider a gradient of 1 in 100 an additional load of 22.4lb is required in each case (Fig 8). This gives the following result:

(a) Total force to pull one ton, for road = 60lb + 22.4lb = 82.4lb
(b) Total force to pull one ton, for railway = 8lb + 22.4lb = 30.4lb.

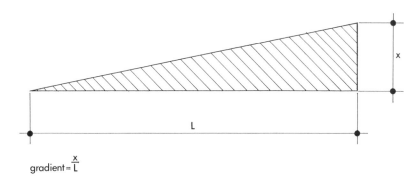

$$\text{gradient} = \frac{x}{L}$$

Fig 7 A gradient

Thus for a gradient of 1 in 100, the additional force required to pull the wagon increases by about one third for the road, but for the railway, nearly four times the force is required, that is, the value increases from 8lb to 30.4lb. In 1840, passenger train weights were about 60 tons and freight trains about 250 tons. Hence a 60-ton passenger train would require 480lb to pull it along a level track but on climbing a gradient of 1 in 100 the force would have to increase to 1,920lb.

One of the most remarkable features of the Great Western Railway was the gentle gradients adopted (Fig 9). For 67 miles of the route the gradient does not exceed 1 in 1,320 and for a further 47 miles it does not exceed 1 in 660, hence the reference to 'Brunel's billiard table'. The easy gradients and gentle curves of the line enabled the Great Western

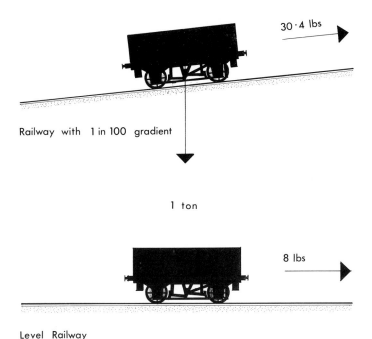

Railway with 1 in 100 gradient

30·4 lbs

1 ton

8 lbs

Level Railway

Fig 8 A large increase in force is required to ascend a gradient of 1 in 100

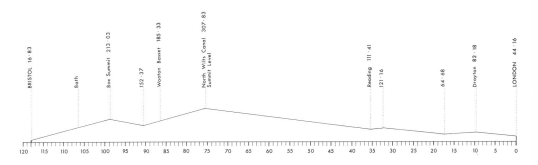

Fig 9 The gradients adopted for the Great Western Railway

Railway, after overcoming initial problems with the track and locomotives, to operate the fastest trains in the world.

In order to obtain the shallow gradients it was necessary to construct embankments, cuttings, viaducts, bridges and tunnels to which must be added the numerous stations, engine sheds, locomotive works, hotels and houses. The first stage of the massive project was to invite tenders for the various sections of the work. Main contractors were then appointed who, in turn, appointed agents for each section of the line. The agents then negotiated with sub-contractors to construct small units of the line under the supervision of foremen or gangers who employed the navvies. The word 'navvy' derives from navigator, which was the name given to the canal builders of the previous century and inherited by railway workers. The way of life of the navvy is vividly described in Terry Coleman's *The Railway Navvies*. The navvy was more than the dictionary definition of a mere labourer – he had a style of his own, distinctive dress and descriptive nicknames such as Fighting Jack, Coffee Joe and Contrary York. The nature of railway construction was such that the navvies lived together in shanties close to the line and then moved on as each section was completed. Their reputation for eating, drinking and wenching was notorious and the Rev D. W. Barnett stated that 'in fact some people have a rough idea that a navvy is a sort of human alligator who feeds on helpless women and timid men, and frightens children into fits'.

The Great Western Railway was constructed largely without the aid of mechanical equipment, and the contractors relied on gangs of navvies working with horses, barrows, picks, shovels and gunpowder. Contractors often took considerable risks, making the accident rate high by modern standards. The death toll during the railway era was many hundreds so it was hardly surprising that navvies occasionally indulged in fights and drunken frolics.

In the construction of the permanent way, Brunel chose to combine the advantages of easy gradients and gentle curves with what is arguably one of the most controversial of his many design decisions – to increase the gauge or distance between the rails from the standard value of 4ft 8½in to 7ft and to support the rails continuously. Table 2 (Whishaw) lists a number of important lines sanctioned prior to the Great Western Railway, all of which adopted the standard or narrow gauge.

LINE	DATE SANCTIONED
Liverpool & Manchester	5 May 1826
Newcastle & Carlisle	22 May 1829
Grand Junction	6 May 1833
London & Birmingham	6 May 1833
London & Greenwich	17 May 1833
London & South Western	25 July 1834
London & Croydon	12 June 1835
GREAT WESTERN	31 August 1835

Table 2 Lines sanctioned prior to the GWR

By the end of 1835, the 7ft 'broad gauge' represented about twelve per cent of the total mileage sanctioned but even Brunel's elegantly presented arguments were unable to prevent eventual defeat in the protracted war of the gauges. However, it was not until 1892 that the last broad gauge trains ran on the Great Western Railway. Following experience of travelling on narrow gauge lines with cramped accommodation and uneven running, and a conviction that speed and smooth running were important to the traveller, Brunel ignored the precedent set by other railway interests and submitted the following observations in a report to the Directors of the Great Western Railway in September 1835:

1 The importance of shallow gradients – for about two thirds of the route, the gradient does not exceed 1 in 1,320 (4ft per mile).
2 Resistance from friction diminishes in the same ratio that the diameter of a wheel is increased.
3 By widening the rails, the body of the carriage could be kept within the wheels and its centre of gravity lowered producing steadier motion and enabling the wheel diameter to be increased.

He then lists the objections – extra land to be acquired, increased friction on curves, heavier carriages, the meeting of the broad and narrow gauges at the junction with the London & Birmingham Railway and the use of a joint London terminus. The last objection was considered by Brunel to be the only one of any real importance and this was resolved for him by the decision of the Great Western board to obtain a separate entry into London.

After detailed consideration, the proposal for a 7ft gauge was adopted and the approximate general dimensions of the permanent way were as follows:

Double line of seven feet	14
Space between the two lines	6
Space outside each line, five feet	10
	—
Total	30ft
	—

For the Liverpool & Manchester Railway, the total width of way for the double line was 25ft 7in and the type of rail and mode of support was completely different from that proposed by Brunel for the Great Western Railway. The Stephensons chose to support fish-belly rails at intervals of 3–5 ft (Fig 10). The supports initially consisted of red sandstone blocks, but due to problems with splitting, they were gradually replaced by half-log sleepers placed at right angles to the rails. For the London &

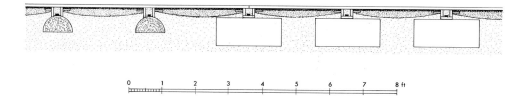

Fig 10 The Stephenson's method of rail support for the Liverpool & Manchester Railway

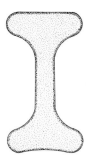

Fig 11 The double parallel rail, London & Birmingham Railway

Birmingham Railway, a double parallel form was adopted (Fig 11). A number of experiments with iron rails of different shapes were made in the first half of the nineteenth century, and initially they were considered as simple beams spanning between fixed supports subjected to moving loads. Subsequently, they were considered as being continuous over a number of supports and towards the end of the century, the theory of beams continuously supported on an elastic foundation (see Technical Appendix) was used to estimate stresses in rails.

Brunel chose to adopt continuous support to the rails which was the oldest form of construction, but there was one important difference – the introduction of piles to hold down the longitudinal timbers supporting the rails. The principle of the construction of the permanent way *c.* 1838 was that longitudinal sections 12–14in by 5–7in and 30ft length are braced at 15ft intervals by cross members as illustrated in Fig 12. A double cross member (transom) is introduced at each end of the 30ft longitudinal members and the piles are connected to the cross members at 15ft intervals. The purpose of the piles was not to support the longitudinal members but to enable ballast to be firmly packed beneath them. A 'bridge' rail (Fig 13) was designed by Brunel to give the best strength to weight ratio and mounted on a tapered hardwood plank on top of the longitudinal members. The bridge rail of length 14–17ft was

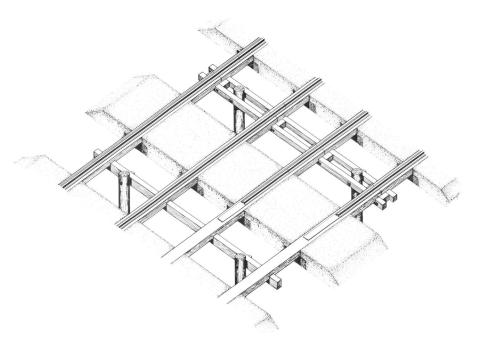

Fig 12 Brunel's design for the permanent way

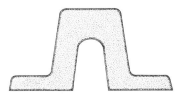

Fig 13 Bridge rail

bolted directly to the longitudinal members and weighed from 45–62lb per yard.

The use of a continuous support to the rails was theoretically correct in terms of minimising stresses and deformation, and a further argument was that the strength of the wrought iron used for the rails could not be relied upon. In practice, the design was a disaster, as the ballast subsided and the longitudinal members tended to deflect between the piles giving uneven running. Brunel was forced to reconsider the design but construction using piles had been carried out as far as Maidenhead. Eventually it was decided to cut off the piles and about 1.35 million cubic feet of American white pine was required for the longitudinal timbers alone, which at current prices represents a cost in excess of that expended for the complete construction.

By the end of June 1840, expenditure on the construction was in excess of £4.5 million, about 2.5 per cent of which covered Brunel's salary, office and supervisory expenses. The opening ceremony for the completion of the line took place on 30 June 1841 and the total cost was in excess of £6 million. The construction time of about 5 years 9 months for the 118 miles of double track compares favourably with that obtained for modern motorways. A 75 mile section of the M4 motorway between London and Bristol took 2 years 5 months to complete. The motorway construction had the advantage of mechanised earth moving equipment and the carriageways can to a large extent follow the contours of the land. The Great Western Railway was built virtually by hand, crossed the River Thames three times, in contrast to the one crossing for the M4 motorway, and most significant of all, Brunel's route to the west included the 3,200yd long Box Tunnel.

A train leaving Paddington on the opening day at 8am is stated to have reached Bristol four hours later and between 1838 and 1845 speeds attained on the line were disappointing. A paper read before the Statistical Society in 1843 (Table 3, Macdermot) gave the average speed of the Great Western Railway as 33mph, 3 mph slower than the Northern & Eastern Railway.

LINE	AVERAGE SPEED mph
Northern & Eastern	36
Great Western	33
Newcastle & North Shields	30
North Midland	29
Birmingham & Derby	29
Midland Counties	28
Chester & Birkenhead	28
London & Birmingham	27
Manchester & Birmingham	25

Table 3 Average speeds exclusive of stoppages

However, with the appointment of Daniel Gooch as Locomotive Superintendent and the arrival of the *North Star*, the first reliable engine (see Chapter 9), the situation improved. By 1845 the Exeter expresses completed the journey from Paddington to Bristol in three hours and were easily the fastest trains in the world. From 1845 to the present day, the Great Western Railway's record for high speeds has been most impressive. Between 1845 and 1848 the 8ft singles (4–2–2) *Great Britain* and *Great Western*, designed by Daniel Gooch, achieved speeds in excess of 70mph, and in 1904 the famous *City of Truro* (4–4–0) reached the magical speed of 100mph at Wellington, Somerset. This world record was achieved due to the efforts of two famous Great Western locomotive engineers, William Dean and George Jackson Churchward. Churchward became Chief Mechanical Engineer in 1902.

Until recent years basic railway technology changed little, but the future is promising with the benefits of high technology being utilised by British Rail's Derby research centre. The 125mph High Speed Train (HST) can be exploited to the full on the straight well-graded route from London to Bristol. In June 1973, the High Speed Train maintained 141mph for more than one mile, and on 10 April 1979, a new world record was established with an average of 111.6mph by an HST on the 09.20 Paddington-Chippenham, a 94 mile run. Table 4 compares journey times from Paddington to Bristol Temple Meads in 1845 and 1979.

DISTANCE MILES	STATION	EXETER EXPRESS (1845) TIME	HST (1979) TIME
0	Paddington	09.30	9.45
53	Didcot	10.43	10.26
77	Swindon	11.20	10.48
107	Bath	12.10	11.15
118	Bristol	12.30	11.32
		3 hours	1 hr 47 min

Table 4 A comparison of journey times 1845 and 1979

There are faster runs with fewer intermediate stops to Bristol Temple Meads (1 hour 27 minutes), and one disadvantage of the HST is that it is too fast to observe station names, stations and other constructions on the route, an unlikely drawback in the 1840s.

The frequency and speed of the Inter-City 125 service requires sophisticated push-button electronic control of trains on a space basis which is in complete contrast to the procedures adopted when the first section of the line opened in 1838. The line was patrolled by the company's police who controlled movements of the trains by a time-interval system, using hand signals and lamps after dark. They also switched the points.

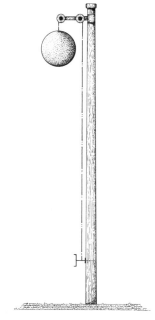

Ball Signal

Fig 14

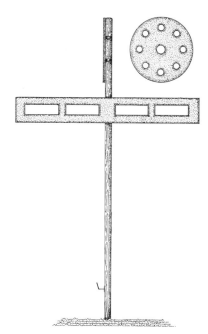

Disc and Crossbar Signal

Fig 15

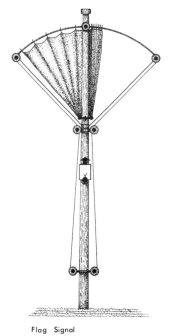

Flag Signal

Fig 16

Fantail Signal

Fig 17

With the speeds anticipated, the need for fixed signals was soon realised, the earliest of which, *c.* 1840, was the signal ball. This consisted of a tall pole up which a large ball could be hauled (Fig 14). With the ball at the top of the pole, the line was clear for the train to move on and if the ball was at the bottom, the train should not pass it. Subsequently, Brunel devised the disc and crossbar signal (Macdermot) which gave a positive indication of all clear and danger. The signal consisted of a timber pole 40 to 60ft high at the top of which was fixed a 4ft diameter disc with an 8ft by 1ft crossbar immediately below it (Fig 15). Both were perforated to reduce wind pressure and the signal could be turned through ninety degrees by means of a lever at the bottom. When the disc faced the direction of travel the line was clear. By turning the signal one quarter of a circle, the flat surface of the crossbar was presented to the driver and this indicated danger and the need to stop. However, there was no means of signifying proceed with caution and to achieve this, Brunel devised a flag signal utilising fabric, but in contrast to the disc and crossbar signal, failed to appreciate that wind would tear the fabric to shreds. This subsidiary signal (Fig 16) comprised of red and green canvas curtains on each side of a post and were attached to rings travelling on a bow-shaped iron rail. Repair costs were excessive and they were soon replaced by fantail signals which were wooden boards in the shape of an arrow painted red on one side and green on the other (Fig 17). The board could be pivoted to give the following instructions:

1 Arrow pointing towards the running line with red side showing – danger.
2 Arrow pointing away from the line with green side showing – caution.
3 Arrow parallel with the line – all clear.

Although Brunel was aware of the development of the electric telegraph by W. F. Cooke and C. Wheatstone, he did not take advantage of its real potential in terms of controlling train movements on a space rather than time interval basis. Cooke and Wheatstone obtained their first patent in 1837, and following disagreement between the directors of the London and Birmingham Railway as to the need for a telegraph, it was installed by the Great Western Railway between Paddington and West Drayton in 1838 and extended to Slough in 1842. Its potential was demonstrated to the public by its assistance in the arrest of a murderer, John Tarvell on 1 January 1845. After administering poison to his mistress in a glass of stout, he travelled by an evening train from Slough to Paddington. The victim's screams were overheard and a message was sent by telegraph to Paddington, enabling the police to make an arrest on the following day.

Brunel's Great Western Railway between Paddington and Bristol can justifiably claim to be one of the most imaginative of all those constructed between 1825 and 1850; it embraces every facet of civil and structural engineering expertise and has great architectural merit. A number of the more important constructions are described in Chapter 3.

3
PADDINGTON TO BRISTOL

It is harder work than I like. I am rarely much under twenty hours a
day at it (I. K. B.)

In order to maintain a thematic division of the subjects, it is necessary
to give separate consideration to four of the most impressive
constructions on the Paddington to Bristol line; that is, Paddington
Station 1838 and 1854 (Chapter 8), the railway village at Swindon
(Chapter 7), Box Tunnel (Chapter 10) and Bristol Temple Meads
(Chapter 8). More detailed technical aspects are covered in Chapter 6,
Bridges, and the Technical Appendix. As we have seen, the works
commenced simultaneously from London and Bristol in September 1835
and the construction was split into nine sections.

SECTION 1: PADDINGTON TO MAIDENHEAD
The line proceeds from Paddington in a westerly direction with Kensal
Green Cemetery, the site of the Brunel family grave, on the right. The
cemetery is, in fact, bordered on the south and north side by the Great
Western and London & Birmingham railways and they are separated by
a distance of about one quarter of a mile. The gradient is approximately
1 in 1,320 and the first major construction, Wharncliffe Viaduct, is
situated some seven miles from Paddington. It is the largest brickwork
construction on the route, 900ft long and the contractors, Messrs
Grissell and Peto commenced work in February 1836. There are eight
semi-elliptical arches 65ft high, 70ft span and rise 17ft 6in. Brunel was
designing arch and suspension bridges simultaneously in the 1830s and
he appreciated that the statical principle was essentially the same (see
Chapter 6), that is, the tension in the chains of a suspension bridge
causing an inward pull, is equivalent to the thrust in an arch rib of
similar (inverted) form which induced an outward thrust on the
supporting piers or abutments. In order to reduce this outward thrust,
the arch was not solid material but of cellular form, the cells being
divided longitudinal walls. The tapering piers which are also hollow have

(Above) An elevational view of Wharncliffe Viaduct which shows Lord Wharncliffe's arms above the central pier on the south face, placed there in recognition of his support for the Act of Incorporation of the Great Western Railway in 1835 *(Frances Gibson Smith)*

(Opposite) The piers on Wharncliffe Viaduct have 'Egyptian' style cornices at the top. The lime mark distinguishes the line of the work undertaken in the 1890s when the viaduct was widened to accommodate four tracks *(Frances Gibson Smith)*

an 'Egyptian style' and the cornice at the top supported the timber centering which carried the brick arches until the mortar had set. The pier foundations are taken down through gravel and river-washed deposits to about three feet into firm blue London clay. The viaduct was originally built to a width of 30ft between the parapets to accommodate two broad gauge tracks but was widened in the 1890s to accommodate four narrow gauge tracks.

The line continues at a gradient of 1 in 1,320 through Slough and on to Taplow, the original site of Maidenhead Station, about one quarter of a mile from the River Thames. The new Maidenhead Station was constructed about thirty years later, one and a half miles west of the

river. Taplow Station marks the end of the first section of the line which was opened in June 1838.

In order to obtain a suitable level for crossing the river at the required headroom for navigation, the line at Taplow is carried on an embankment and a short distance beyond the station, the river crossing is achieved by a bridge that is visually the most pleasing and technically the most daring of all Brunel's designs in brickwork. The main arches, possibly the largest ever constructed in brickwork, are semi-elliptical in form with a span of 128ft and a rise of 24ft 3in. The technical aspects of arch construction are given detailed consideration in Chapter 6 and in the Technical Appendix, but at this stage it is sufficient to state that the load on the arch induces a thrust C (Fig 18) which can be resolved into vertical and horizontal components which are resisted at the supporting abutments. The value of the horizontal component H, causing outward movement at the abutments, is related to the ratio of the span L to the rise h of the arch. The horizontal thrust increases with increase in the ratio L/h, which for the Maidenhead Bridge is about 5.28. This is a high value for a brick or masonry arch especially when compared with the figure of 2.0 for the Roman semi-circular arch.

The main arches are flanked on each bank by four semi-circular flood arches, and in the interior of the first flood arch on each bank, the void between the spandrel walls was filled with concrete to create a thrust opposed to that of the main arches.

In the spring of 1838, the centering supporting the arches was eased and to the delight of Brunel's critics the eastern arch showed signs of movement. This consisted of a separation of about half an inch between

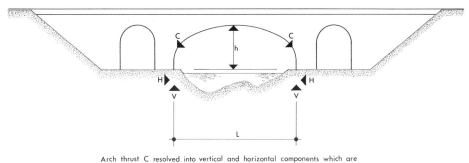

Arch thrust C resolved into vertical and horizontal components which are resisted by the supporting abutments

Fig 18 The resolution of the arch thrust into vertical and horizontal components

The two semi-elliptical main arches of the bridge over the Thames at Maidenhead (*Frances Gibson Smith*)

the lowest three courses of bricks extending about 12ft on each side of the crown of the arch. It was caused by the centering having been eased before the mortar linking the brick courses had properly set, and the contractor admitted liability. The necessary repair work was carried out and the centres eased, but left in position. Brunel ordered that they should not be removed until another winter had passed, but in the autumn of 1839 a violent storm blew them away. The bridge, needless to say, has been standing majestically ever since, supporting with ease the high speed trains of the 1970s. The tracks were quadrupled between 1890 and 1893 which involved widening the bridge, and the junction of the new and old works is clearly shown in a view taken of the soffit of the eastern arch.

The soffit of the eastern main arch at Maidenhead showing the new and old work (*Frances Gibson Smith*)

SECTION 2: MAIDENHEAD TO TWYFORD

After crossing the Thames, the line continues at a gradient of 1 in 1,320 to the end of the second section of the route at Twyford, about thirty-one miles from London. There are no significant structures on this section and the works proceeded smoothly. The first portion is carried on an embankment and after leaving the town of Maidenhead, passes through a series of cuttings in which chalk appears. Between July 1839 and March 1840 Twyford was the terminus of the railway from London as major problems were encountered on the remaining few miles to Reading.

SECTION 3: TWYFORD TO READING

To maintain the gradient at 1 in 1,320 it was necessary to construct the cutting at Sonning, nearly two miles long at a depth varying from 20–60ft through sand, clay and deep water. Fig 19 shows a section through the cutting at its deepest point which required the removal of

7,800 cubic feet of material for each foot of length. The excavation fell behind schedule and at one stage over 1,200 men and 200 horses were employed on the work. They were able to excavate and carry away in excess of 220,000 cubic feet of material per week and this figure was increased by forty per cent with the assistance of two locomotive engines. Had the line been two tracks of narrow gauge the volume of excavation would have been reduced by about six per cent, indicating that Brunel's broad gauge was not too extravagant in terms of use of land. Section 3 was not opened until the end of March 1840 and at its deepest point is crossed by two bridges. The photograph of the cutting shows an iron bridge which spans the four narrow gauge tracks. It was constructed in 1893 to replace a most significant timber bridge designed by Brunel to span the two broad gauge tracks.

The timber bridge can be considered as the embryo for a large number of timber structures constructed some years later between Exeter and Penzance. The bridge had an overall span of about 240ft and consisted of four tapering timber piers from which raking struts fan out to support the deck and reduce its effective span. It was designed for light traffic but structures of this form were later used to carry train loads of over 200 tons.

A few hundred yards to the west of the iron arch bridge, the London to Reading turnpike road crossed the line and is supported by a three span brick arch bridge. This now forms the A4 trunk road, one carriageway being supported by the original brick arch and the other, immediately to

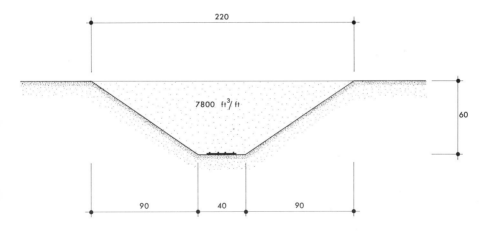

SONNING CUTTING   volume of excavation at deepest point

Fig 19

The two mile-long cutting at Sonning (*University of Surrey*)

The timber bridge crossing Sonning Cutting designed by Brunel, in a lithograph by J. C. Bourne

the west, by a recent concrete structure. On leaving Sonning Cutting, the line meets the Thames valley, leading to the end of section 3 at Reading Station, some thirty-six miles from London. Reading Station, constructed on an embankment, was again of controversial design, with the up and down platforms in line, a short distance apart, on the south side of the railway. The town lay almost entirely on the south side of the railway and Brunel argued that it was a very convenient arrangement for passengers. Each platform had its separate loop and through trains ran clear of the station. However, there were obvious junction problems associated with the connection of the loops to the through lines (see Chapter 8). On Monday 30 March 1840, the line to Reading was opened for public traffic.

SECTION 4: READING TO STEVENTON
For well over half of the twenty miles from Reading to Steventon, the line follows closely the route of the Thames via Pangbourne and crosses it twice, at Basildon and then at Moulsford. At Moulsford, the line departs from the Thames and takes a north westerly direction on to Steventon. Brunel's multi-span brick arch bridges at Basildon and Moulsford, although not as spectacular as the shallow arch bridge at Maidenhead, have an artistic merit which complements one of the most peaceful stretches of the river with wooded banks and extensive meadows. They are best viewed from the towpath and access is not easy, but well worth the effort.

(*Above*) Moulsford Bridge taken from the east bank (*University of Surrey*)

(*Opposite*) Moulsford Bridge – the gap between the new and the old work (*University of Surrey*)

Basildon Bridge was widened to accommodate the quadrupling of the track, and the junction of the old and new work can be seen, the skew being about fifteen degrees. Two miles upstream from Basildon Bridge, Moulsford Bridge crosses the Thames at a skew angle of forty-five degrees and the elevational view is taken from the towpath on the east bank. The gap between the new and the old work may be seen in the photograph and a group of islands can just be seen in the bottom right-hand corner. The riverside sounds of insects, birds and water lapping against the banks are only slightly marred, every fifteen minutes or so, by the purr of the high speed trains as they wind their way along the valley at average speeds in excess of 100mph. The journey time from Reading to Moulsford Bridge is under seven minutes and this emphasises the view that the High Speed Train is not the best means of observing the structural and aesthetic merit of the railway as it passes through this unique stretch of the Thames valley.

A few hundred yards west of Moulsford Bridge the line is crossed by the Reading to Wallingford road (A329) and this is a convenient point for observing the change from embankment to cutting. A view looking south-east towards Moulsford bridge shows Brunel's hotel for Wallingford Road Station in the top left-hand corner. This station was superseded by Cholsey and Moulsford station in 1892. Travelling north-west, the line enters a chalk cutting as it curves round to Cholsey Station, and then on to Didcot and Steventon, fifty-six miles from London. The original station at Didcot and the line to Oxford were not opened until June 1844.

The High Speed Train covers the distance from Reading to Didcot in fourteen minutes, whereas the stopping trains on the Paddington to Oxford service take thirty minutes. This service presents a much better opportunity to observe the line, and a few miles to the north of Didcot at Culham there is probably the only surviving example in public service of a wayside station designed by Brunel (see Chapter 8). For a short period of time, Steventon achieved the status of a terminus and trains were in

Looking south-east towards Moulsford Bridge with Brunel's hotel for Wallingford Road Station on the left *(University of Surrey)*

House at Steventon used as the weekly venue of the board of GWR (*Derrick Beckett*)

operation from 1 June 1840. The District Superintendent's house, *c.* 1839 with Brunelian characteristics, was recently advertised for sale at £48,500 and had the distinction of being the weekly venue for the Board of the Great Western Railway (Fuller) when the separate London and Bristol committees were combined. Steventon was a convenient location as it is near the midpoint of the line and meetings were held there during the latter half of 1842. From January 1843, board meetings were held in London but the station was still important as it maintained a frequent coach service to Oxford. It was closed to passenger traffic in 1964 and the station buildings subsequently demolished and the goods yard cleared. However, the house remains.

SECTION 5: STEVENTON TO FARINGDON ROAD (CHALLOW)
This section of the line was completed seven weeks after the opening to Steventon and there were no major structures involved. The seven miles are almost straight and the line rises from east to west at a gradient of about 1 in 750 via shallow cuttings and embankments. The village of Faringdon is situated a few miles to the north of the line and the station, subsequently renamed Challow, no longer exists.

SECTION 7: FARINGDON ROAD (CHALLOW) TO HAY LANE

In order to maintain continuity it is necessary to deviate from the date order of completion (see page 37), and section 7 takes the line on a further sixteen miles to a distance of just over eighty miles from London and about three miles from the old town of Swindon. At the time of the completion of section 7 of the line on 17 December 1840, the new town and railway works (see Chapter 7) were only in the planning stage. The M4 motorway now crosses the railway about two miles west of Swindon station and a short distance after, a minor road (Hay Lane) also crosses the line at the point where it enters a cutting, which extends for about one mile. Hay Lane acted as a temporary terminus until the opening of section 8.

SECTION 8: HAY LANE TO CHIPPENHAM

This section of the line, completed on 31 May 1841, starts at the entrance to the long cutting at the junction with Hay Lane. Skirting to the south of Wootton Bassett (now junction with the line to Bristol Parkway and South Wales), the line, between mileposts 85 and 87 falls from east to west at a gradient of 1 in 100, known as the Wootton Bassett incline. Beyond this point the line reverts to a gradient of 1 in 660. The line approaches Chippenham Station via an embankment; the station forebuilding, but not the platform structures, is a Brunel design (see Chapter 8).

SECTION 9: CHIPPENHAM TO BATH

In the thirteen miles from Chippenham to Bath, the last section to be completed, Brunel was faced with a number of major engineering problems. Immediately to the west of Chippenham Station, ninety-four miles from London, the junction of the A4 and A420 roads is straddled by one of the least inspiring of Brunel's designs. A massive arch bridge, main span 26ft with two side spans of 10ft over the footpaths, is followed by a six arch viaduct. The structures were originally built in Bath stone, but with the passage of time, there has been extensive patching up with brickwork. The viaduct leads on to an embankment which extends for two miles. This is followed by almost continuous cutting for a distance of about three miles, the latter part of which is cut with perpendicular sides through the Bath stone.

(*Opposite*) An HST leaving the west portal of Box Tunnel (*University of Surrey*)

Just beyond milepost 99, the line enters the 9,600ft-long Box Tunnel, which at the time of construction was by far the longest railway tunnel in the country. At the Lords Committee stage, Brunel's proposal to drive a tunnel under Box Hill at a gradient of 1 in 100 met with fierce opposition, in particular from a pontificator of pseudo-science, Dr Dionysius Lardner. He proved by means of elaborate calculations that if the brakes were to fail as a train entered the tunnel on a falling gradient it would emerge at the west portal at a speed of 120mph, a speed, he added, at which no passenger would be able to breathe. Brunel demolished his arguments by observing that factors such as friction and air resistance must have been lost in the doctor's calculations, because, owing to their combined effect, the speed would be 56mph. Although weathered, the western portal, depicted in a photograph, still forms an impressive entrance to one of the most hazardous civil engineering constructions that Brunel encountered. A more detailed technical description of the tunnel is given in Chapter 9. It is believed that Brunel so aligned the tunnel that on his birthday, 9 April, the sun briefly shines through it.

The line leaves the west end of the tunnel at milepost 101.25, passes under the A4, continues on an embankment leading to Middlehill Tunnel. From the 630ft-long tunnel, it continues on an embankment, crossing the River Avon, and enters the eastern approaches to Bath through Sydney Gardens. This involved the construction of a number of bridges and retaining walls which enhance the architecture of this elegant city. The photograph taken in Sydney Gardens, shows a cast-iron arch footbridge designed by Brunel. The span is 30ft and at one side it is supported by the curved retaining wall of the cutting. It was found necessary to divert the Kennet & Avon Canal, and the retaining wall was required to support both the cutting and the canal. Immediately to the east of Bath Station the River Avon is crossed by St James Bridge which is an elliptical arch of 88ft span and 28ft rise.

SECTION 6: BATH TO BRISTOL

At the west end of Bath Station the line again crosses the River Avon at a considerable skew and is supported on a two span trussed steel girder bridge. This is the site of an early design by Brunel for a timber bridge. It is described in J. C. Bourne's history of the Great Western Railway:

The bridge is of two arches, each of 80ft span. Each arch is composed of six ribs placed about 5ft apart and springing from the abutment and from a central masonry pier. Each rib is constructed of five horizontal layers of Memel timber held together by bolts and iron straps. The end of each rib is enclosed in a shoe

IT IS A REMARKABLE FACT THAT ANNUALLY ON THE MORNING OF APRIL 9TH, THE SUN'S RAYS PENETRATE THROUGH THE GREAT BOX TUNNEL OF THE GREAT WESTERN RAILWAY AND ON NO OTHER DAY IN THE YEAR.

THE DAILY TELEGRAPH APRIL 12TH 1859

EVEN MORE REMARKABLE IS THE FACT THAT APRIL 9TH IS THE BIRTHDAY OF BRUNEL.

The BOX TUNNEL
BUILT BY
I.K.BRUNEL

COMMENCED 1836
OPENED
JUNE 30TH 1841

or socket of cast iron, resting, with the intervention of a plate, upon the springing stones, the shoes on the middle pier being common to the two ribs. The spandrils of the four external ribs are filled up with an ornamental framework of cast iron supporting the parapet. The interior ribs are connected by cross struts and ties.

The wooden bridge was replaced in 1878 which, in turn, has been replaced by the bridge shown in the recent photograph. Between Bath and Bristol the gradient is generally 1 in 1,320, falling from east to west. Westward from Bath Station, the line is carried on a long viaduct which was originally stone-faced but this has been largely replaced by brick. From the suburbs of Bath, the railway and A4 road run approximately parallel, and this section of the line, approximately eleven miles in length, took over four years to complete, the works including tunnels, viaducts, embankments, retaining walls and bridges. The terminus at Temple Meads comprised of a passenger station and goods depot and communicated with the terminus of the Bristol & Exeter Railway. This terminus was situated at right angles to that of the Great Western and

(*Opposite*) Cast-iron footbridge at Sydney Gardens in Bath (*Frances Gibson Smith*)

(*Below*) The timber arch bridge over the Avon at Bath in a lithograph by J. C. Bourne

A two-span trussed steel girder bridge on the site of Brunel's earlier laminated timber arch bridge (*University of Surrey*)

was not completed until 1845. The Great Western passenger station, now used as a car park, will be described in Chapter 8. The rail network at Bristol was further complicated by the completion of the Bristol & Gloucester Railway in July 1844 and thus for a period of time there were three railways sharing two stations. Less than eight years after the opening of the line from London to Bristol, the link with Exeter and Plymouth had been completed.

4
BRISTOL TO PLYMOUTH

BRISTOL TO EXETER

A bill for the construction of a railway between Bristol and Exeter, based on plans drawn up under the supervision of Brunel, became an Act of Parliament in May 1836. The Bristol & Exeter Railway Company was authorised to construct the seventy-six mile line and committed itself to the broad gauge. Initially, due to financial difficulties, progress was slow, and by the end of 1839 it was proposed to lease the line to the Great Western in order to avoid the cost of purchasing engines and rolling stock. This proposal received Parliamentary assent and the leasing arrangement was to continue until five years after the completion of the whole line to Exeter. Finances improved and by the end of May 1841, thirty-three miles of double track were completed between Bristol and Bridgwater. There was also a single track branch line to Weston-super-Mare. The loop line to Weston-super-Mare was not opened until 1884.

Apart from a number of bridges and Whiteball Tunnel, there are no major engineering works between Bristol and Exeter but about one mile south of Bridgwater, Brunel was responsible for the design of a bridge crossing the River Parrett. The bridge was of masonry arch construction with a span of 100ft and a rise of 12ft. The span to rise ratio of 8.33 is large for this type of construction and considerably in excess of that adopted for the Thames bridge at Maidenhead. Due to movement of the foundations Brunel chose to maintain the centering in position but this obstructed the river and he was eventually forced to substitute a timber arch between the same abutments. The timber arch was replaced in 1904 by the steel girder bridge shown in the photograph. An inspection of the river bed at low tide indicates soft bearing strata and this was probably the cause of the foundation movement in the original masonry arch.

Another interesting example of bridge design adopted by Brunel on the Bristol and Exeter line is the 'flying bridge' in which the arch springs from the sides of the cutting which it helps in some measure to support. A major example of this type of construction is located in a cutting near Bleadon to the south of Weston-super-Mare. The cutting reaches a depth

The present bridge over the River Parrett. Brunel's original masonry arch structure underwent excessive settlement (*Derrick Beckett*)

of 60ft and the flying arch spans 110ft. To the south of the cutting the line runs in a straight line, almost level to Bridgwater.

By the spring of 1844, the 3,280ft-long Whiteball Tunnel on the Somerset and Devon border was completed. The tunnel was constructed on a gradient of 1 in 128 and due to the poor soil conditions, it is lined throughout in brickwork.

The railway from Bristol to Exeter was opened on 1 May 1844 (Rolt) thus completing the 194 miles of broad gauge track between Paddington and Exeter. On the opening day, Daniel Gooch was able to demonstrate the speed and reliability of the Fire Fly class engine, the first he designed. He worked a special train with six carriages, hauled by the 24-ton 2–2–2 *Actaeon*, which departed from Paddington at 7.30am. The train carried a large party of dignitaries, including a member of Parliament, Sir Thomas Acland. The journey took five hours and after a dinner in the goods shed at Exeter Station, the party left at 5.20pm arriving at Paddington Station four hours forty minutes later. At 10.30pm, Sir Thomas Acland was able to tell the House that he had been in Exeter at 5.20pm. The combination of Gooch's locomotive design and Brunel's broad gauge track enabled the Great Western, in the following year, to operate the fastest scheduled train services in the world – the Exeter expresses (see Chapter 2).

EXETER TO PLYMOUTH

Between 1836 and 1844 a number of proposals were put forward for a line to connect Exeter with Plymouth and an Act for its construction received Royal Assent in July 1844. The route chosen for the South Devon Railway (Fig 20) follows the west bank of the River Exe via Starcross, hugging the coastline and passing through Dawlish and Teignmouth. At Teignmouth it turns inland, running parallel to the east bank of the River Teign and on to Newton Abbot. From Newton Abbot it climbs steeply, reaching a summit at Dainton Tunnel, and then continues on a falling gradient to Totnes. Between Totnes and Rattery, the line rises at an average gradient of 1 in 64, passes through Marley Tunnel, 2,610ft long, and continues to climb to a summit at the junction with the Kingsbridge road near Wrangaton. The line then descends into

Fig 20

Plymouth via Ivybridge and Plympton. The average gradients between Newton Abbot and Plymouth are shown in Fig 21 (Macdermot) and it is of interest to compare them with those used between London, Bristol and Exeter. For financial reasons, Brunel was forced to adopt sharper curves and steeper gradients for more than sixty per cent of the fifty-two mile South Devon Railway. The first section, between Exeter and Newton Abbot, is almost level but the imponderable nature of the force of the sea was to be contended with. This problem was not new to Brunel, as by 1844 he had successfully designed and launched two of the world's largest ships, the PS *Great Western* and the SS *Great Britain*.

Thus the South Devon Railway posed a number of exacting construction problems including sea walls, tunnels, sharp curves, steep gradients and long viaducts. Further, Brunel wished to maintain his principle that speed within reasonable limits is a material ingredient of perfection in travelling. Brunel's reservations about the ability of locomotives to cope with the steep gradients between Newton Abbot and Plymouth led to his involvement in a project which was known to the locals as the 'Atmospheric Caper'.

It was indicated in Chapter 2 that the force required to pull a load of one ton on a level railway is 8lb. With the proposed gradients of 1 in 40, the additional load to pull one ton is 56lb, that is seven times that required on the level. As the locomotives in use at the time weighed about 25 tons and formed a large portion of the total train load, it is not unnatural that Brunel's thoughts should turn to another form of traction – the atmospheric system.

The atmospheric system, patented in 1839 by Mr Samuel Clegg and

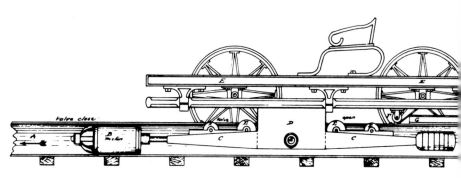

Fig 22

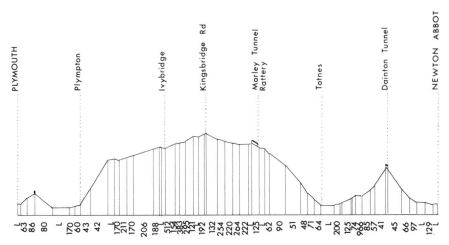

Fig 21 Average gradients between Newton Abbot and Plymouth

Messrs Jacob and Joseph Samuda, underwent trials on an experimental section laid down at Wormwood Scrubs in 1840 and attracted the interest of a number of engineers including Robert Stephenson and Isambard Kingdom Brunel. It was applied on the Dalkey extension of the Dublin & Kingstown Railway and on the London and Croydon line. While Robert Stephenson rejected the system on economic grounds, Brunel presented a detailed report to the directors of the South Devon Railway in August 1844 which concluded as follows: 'I have no hesitation in taking upon myself the full and entire responsibility of recommending the adoption of the Atmospheric System on the South Devon Railway, and of recommending as a consequence that the line and works should be

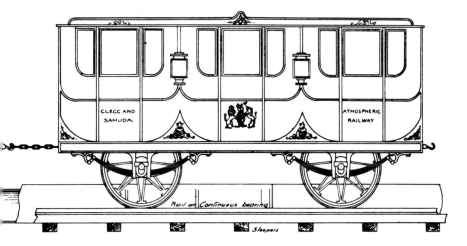

constructed for a single line only.' The report presents convincing arguments for the adoption of the system for reasons of economy and reduced journey times.

Briefly the principle of the system as used on the SDR is as follows: in Fig 22, a cast iron pipe (A), 15 inches in diameter with a continuous slot at the top, is laid between the rails. The piston (B) is connected by means of plates (C) and (D) to the railway carriage framing (E). The continuous slot at the top of the pipe is sealed along its length by means of a reinforced leather flap valve sealed by means of grease to make it airtight. At intervals along the line, stationary engines were erected and worked air pumps which created a partial vacuum in the slotted pipe. In theory, assuming no losses due to leakage, it is possible to develop a complete vacuum in front of the piston (Fig 23). The pressure of the atmosphere at sea level is in the order of 14.7lb per square inch and so this pressure difference between the front and rear of the piston is equivalent to a force of 1.16 tons.

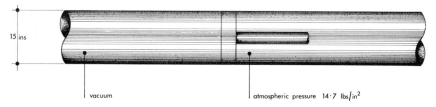

15 ins

vacuum   atmospheric pressure 14·7 lbs/in²

Fig 23 Force created by a vacuum

The directors accepted Brunel's report, arrangements were made with the Samuda Brothers for the use of their patent rights and the tenders accepted for the manufacture of the steam engines and vacuum pumps were those of the illustrious names Boulton & Watt, G. & J. Rennie, and Maudslay Son & Field. Brunel hoped that the line between Exeter and Newton Abbot would be operational in the summer of 1845 but damage to the sea wall and delays in the construction of the tunnels in the neighbourhood of Dawlish put the work far behind schedule and it was not until 30 May 1846 that the fifteen miles between Exeter and Teignmouth opened for passenger traffic only. There was little progress with the atmospheric system and initially engines were hired from the Great Western Railway. It was not until September 1847 that atmospheric trains carried passengers between Exeter and Teignmouth and this service was extended to Newton Abbot in January 1848.

(*Opposite*) Relics of the atmospheric system at Starcross

However, there were a number of technical problems — rapid deterioration of the leather flap valve, leakage of air in the pipe, frequent breakdowns at the engine houses and excessive fuel costs. Brunel was forced to admit defeat and after 9 September 1848 the line was worked throughout by locomotives. It is significant to note that the highest speed recorded with the atmospheric system was 68mph with a train of 28 tons running on a level track.

Fortunately a number of relics of the atmospheric system remain, including the engine house at Starcross, built in the style of an Italian bell tower, together with the sea wall and the Atmospheric Railway public house. Regrettably, the structure of the engine house has been allowed to deteriorate and restoration would seem desirable as it is an excellent example of Brunel's frequent use of an architectural style to disguise the true function of a building. This suggests a lack of originality and certainly would not find favour in 'progressive' architectural circles. However, there is enough originality in Brunel's timber viaducts west of Plymouth to excuse this dalliance with Italianate façades.

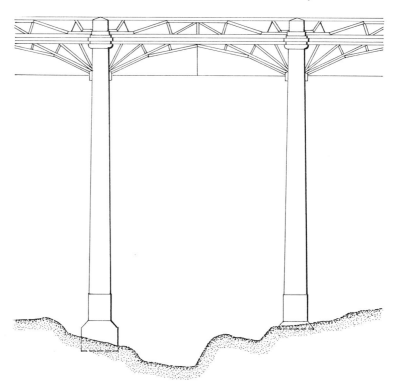

Fig 24 Ivybridge Viaduct

The course of the line between Newton Abbot and Plymouth takes it over the southern slopes of Dartmoor and it was necessary to cross a number of steep valleys. Viaducts with timber frameworks supported on masonry piers were constructed at Glazenbrook, Bittaford, Ivybridge, Blatchford and Slade. The viaduct at Ivybridge has eleven spans of 60ft with a maximum pier height of 104ft. It is illustrated in Fig 24 and described in *The Life of Isambard Kingdom Brunel* as follows:

The piers are of masonry each consisting of two slender and slightly tapered shafts about 7ft square rising to the level of the rails. The superstructure was originally designed for a railway on the atmospheric system and was therefore only intended to bear the load of a train of carriages. The framework was placed below the level of the rails and it consists of a polygonal frame with a few subsidiary struts, the feet of the main timbers being tied together by wrought iron rods.

Prior to construction of the viaduct a complete span of the superstructure was erected and test loaded at Bristol. As the atmospheric system was abandoned, it was necessary to strengthen the structure to support the weight of locomotives. This was achieved by adding a trussed timber parapet above the main frames. Some years later wrought-iron girders were inserted between the timber framing.

It was not until April 1849 that trains were able to work the 264-mile run from Paddington to Plymouth via Bristol and Exeter with a journey time of 7 hours 5 minutes. In October 1979 British Rail commenced High Speed Train services from Paddington to Plymouth taking the shorter route, leaving the Paddington to Bristol line at Reading and joining the Bristol-Exeter line at Cogload Junction which is north of Taunton. A typical journey time for the shorter route is 3 hours 39 minutes.

5
PLYMOUTH TO PENZANCE

The opening of the line from Paddington to Plymouth was an achievement which no doubt brought great satisfaction to Marc Isambard and Isambard Kingdom Brunel. The elder Brunel, fifty years previously, had commenced a wearisome coach journey from Plymouth to London which lasted several days and now his son had achieved the remarkable feat of developing a transport system which reduced the journey time to a little over seven hours. Marc Isambard died in December 1849 and thus was unable to enjoy his son's progress with the extension of the line from Plymouth to Penzance.

The works included the Royal Albert Bridge at Saltash, which is about three miles north-west of Plymouth and the construction of the bridge is described in Chapter 6. An Act for the construction of a railway from Plymouth to Falmouth with a number of branches received Royal Assent in August 1846. The Cornwall Railway Company appointed Brunel as the Engineer and its capital was supplemented by injections from the Great Western, Bristol & Exeter, and South Devon railways. Due to financial restrictions little work was carried out until 1853. By this time contracts had been let for the Royal Albert Bridge and for most of the line, but it was not until May 1859, a few months before Brunel's death, that passenger trains ran from Plymouth to Truro. Work was always hampered by lack of funds and the construction problems confronting Brunel, associated with the sharp curves, steep gradients and deep valleys were compounded by the additional work load of providing the engineering expertise for the construction of the 2,200ft long Royal Albert Bridge and the 629ft long iron steamship, the PSS *Great Eastern*.

The principle construction problem on the Cornwall Railway was the need to evolve economic designs for the forty-two viaducts between Plymouth and Falmouth, thirty-four on the Plymouth to Truro section, opened in 1859, and eight between Truro and Falmouth, opened in 1863. The total length of viaduct structures was in the order of four miles. The highest of these viaducts exceeded 150ft and many of them were built on steep gradients and sharp curves. Due to

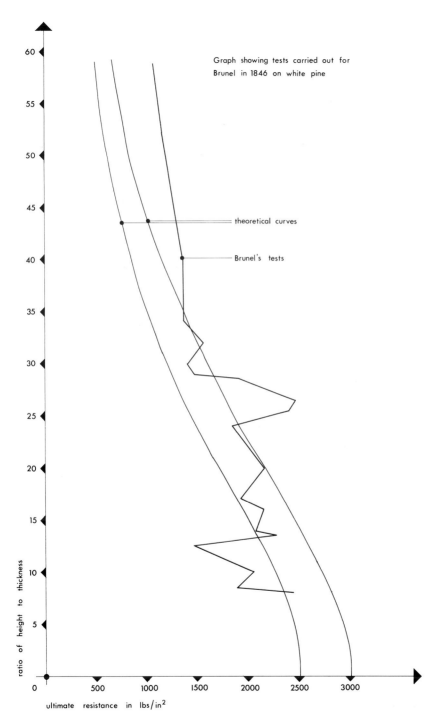

Fig 25 The results of strength tests carried out by Brunel in 1846 on timber columns

severe restrictions in capital available, Brunel chose to design light timber structures to replace, as far as possible, masonry and wrought iron. It will be recalled that some twenty years earlier, he had designed a light timber bridge to span Sonning Cutting (see page 85) and this was the basis for a number of interesting viaduct designs which were adopted on the Cornwall Railway. Brunel's sketch and calculation books at Bristol University Library reveal how he developed ideas of using triangulation and raking struts to reduce deck spans, and some original drawings remain in the archives at Paddington Station. Regrettably, the last of these timber viaducts was replaced in 1934. The principal dimensions of the timber works between Plymouth and Penzance have been tabulated by E. T. Macdermot and L. G. Booth, and details of maintenance procedures and their eventual demolition were recorded by H. S. B. Whitley in 1931.

In using timber for structures carrying train loads in which the engine weight was in the order of 25 tons, Brunel was faced with a number of problems. He required data on strength properties, formulae for strength and deflection, preservation treatments and suitable connection devices to transfer loads from one element to another. Some test data was in fact available to Brunel but he also carried out his own tests. The graph shown in Fig 25 illustrates some erratic results obtained by Brunel in 1846 from experiments which related the compressive strength (see Chapter 6) of timber columns to the ratio of their height to thickness. The specimens were referred to as yellow pine timber and range from 6in to 15in square. Superimposed on these results is a theoretical curve (see Technical Appendix).

There was of course the necessity to protect the structures against fire and biodeterioration, and the process of Kyanising was used as a preservative measure and as a means of making the timber less inflammable. This consisted of placing thoroughly dried members in an open tank which was filled with perchloride of mercury and leaving them in it for a time equal to one day per inch of the widest side of the section. The penetration amounted to about one quarter of an inch. As a further precaution a water tank was placed at each end of the viaducts.

The original timber used was probably obtained from the Baltic and was high quality close-grained pine cut from large trees of slow growth. The quality of the timber and the large cross sectional dimensions and lengths available are a source of envy to the modern timber engineer who leans heavily on technology to obtain optimum performance from much inferior material. It must be admitted that a number of the viaducts were under designed, but the main problem was biodeterioration. Brunel was aware that yellow pine was not a species of unlimited natural durability and that the preservation techniques were questionable; he therefore designed the structures so that, in general, defective members would be replaced with minimum interruption to traffic. After World War I yellow pine became scarce and the replacement material was of limited durability which led to uneconomic maintenance costs.

Turning to the structural forms adopted for the Cornwall Railway viaducts, they can be roughly classified as follows:

1 Inclined legs, two sets of three, springing from small masonry piers. The legs supported three longitudinal laminated beams constructed from two 24in by 10in pieces of timber connected by bolts and joggles. This form of construction was used as an alternative to an embankment in shallow valleys. A typical example is the Penadlake Viaduct, milepost 272, about two miles east of Bodmin Road station, which is illustrated in diagrammatic form in Fig 26. The span between the piers is 40ft and this structure has been analysed by a final year Civil Engineering Student (C. A. Mercer) at the University of Surrey. A summary of the analytical results and test data is included in the Technical Appendix. Of particular interest, is Brunel's attempt to laminate the two longitudinal members to form a single member with strength properties in excess of those based on the addition of the separate contributions of the two laminates.

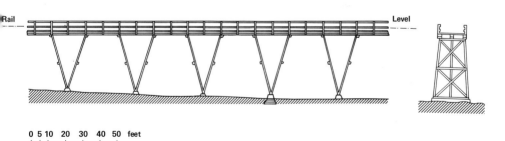

0 5 10 20 30 40 50 feet

Fig 26 Penadlake Viaduct (*University of Surrey*)

2 For deeper valleys, three sets of legs were used springing 41ft below the deck. It was then possible to increase the spans to 50ft and Angarrack Viaduct, milepost 317, is typical of this form of construction. It is on the West Cornwall line and was constructed in 1852. The structural scheme is illustrated in Fig 27 and subsequently the timber piers were replaced by masonry.

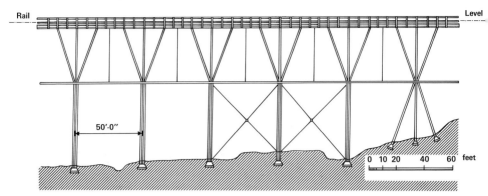

Fig 27 Angarrack Viaduct (*University of Surrey*)

3 Four sets of legs springing from masonry piers which were at 60–66ft centres. St Pinnock Viaduct (see Fig 28) milepost 269, illustrates this form of construction. The maximum pier height was 151ft and it was constructed in 1859.

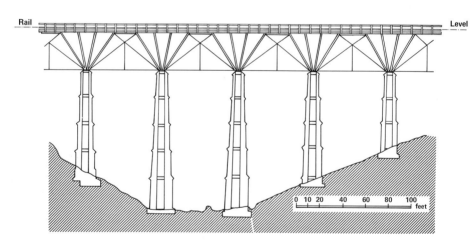

Fig 28 St Pinnock Viaduct (*University of Surrey*)

4 Fig 29 illustrates an example of use of timber piers where the line crossed tidal creeks. The spans were generally reduced to about 40ft but in this case, the St Germans Viaduct, milepost 256, the span was increased to 66ft and the raking member construction was replaced by a timber and wrought-iron truss.

Mining interests in the Redruth and Hayle area of west Cornwall led to the construction of a narrow gauge single line twenty years before the opening of the Cornwall Railway in 1859. An Act for the construction of

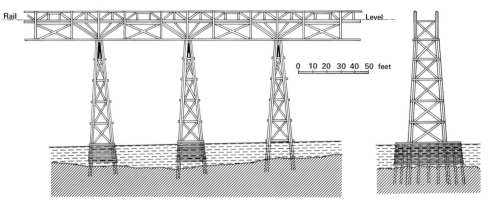

Fig 29 St Germans Viaduct (*University of Surrey*)

a broad gauge railway between Truro and Penzance received Royal Assent in August 1846 and it was known as the West Cornwall Railway. Again, lack of capital limited progress, and to construct the line as cheaply as possible it was decided in 1850 to adopt the narrow gauge which avoided the expense of altering the original mineral line and its rolling stock. However, the constructions were made wide enough to accept the broad gauge when funds became available.

Additional economies were made by Brunel's adoption of a rail designed by William Barlow, Engineer of the Midland Railway Company. The Barlow rail (Fig 30) was laid directly on the ballast, thus

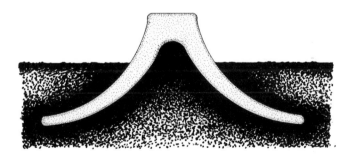

Fig 30 The Barlow rail laid directly on ballast

avoiding the use of supporting timbers, but was found to be unsatisfactory for heavy traffic. The 25 mile single track line from Truro to Penzance was opened in August 1852 and there were nine timber viaducts, including Angarrack which has been described. It was not until several years after Brunel's death that it was decided to lay broad gauge track on the West Cornwall Railway. The work involved strengthening the timber viaducts to carry heavier trains. In March 1867, the first

The original piers of Moorswater Viaduct (*Ian Barnes*)

through passenger service on the 327-mile broad gauge line from
Paddington to Penzance was inaugurated and the fastest journey time
was twelve hours.

What is now left of the structures conceived by Brunel for the South
Devon, Cornwall and West Cornwall railways? As we have seen, the
timber deck structures to the viaducts were removed by 1934 but a
number of examples remain of the masonry constructions. Typical of
these are the masonry piers of the two viaducts immediately to the west

and east of Liskeard Station, milepost 265. Moorswater Viaduct, to the west, has a maximum pier height of 147ft and the photograph shows the original piers built in the form of a cross which thickened in steps according to the height. The original viaduct was replaced in 1881 and the new structure can be seen in the background. Liskeard Viaduct, to the east has a maximum pier height of 150ft and in this case the original piers were extended, as with the St Pinnock Viaduct to support a shallower iron truss. In the photograph of the piers of St Pinnock Viaduct, milepost 269, the new masonry between the underside of the truss and the top of the original construction, in cross form, can be

St Pinnocks Viaduct (*Ian Barnes*)

The flying arch at Liskeard Station (*Ian Barnes*)

clearly seen. This was the highest viaduct on the line with a maximum
pier height of 151ft. It was replaced in 1880.

On a smaller scale there is an interesting masonry arch which carries
the road over the railway at Liskeard Station. It is referred to as a 'flying
bridge' and instead of the arch being supported on piers and abutments,
it springs from the slopes of the cutting. This type of arch bridge can be
constructed before the cutting is excavated to its full depth, thus avoiding

the use of centering and reducing the volume of masonry required. Another example of a flying bridge is on the Bristol & Exeter Railway near milepost 138.

At the time of the opening of the Cornwall Railway, Brunel's railway interests extended to Land's End, South Wales, and the Midlands with a total track length in excess of 1,200 miles.

6
BRIDGES

Let me give you one general piece of advice. . . . Consider all structures, and all bodies, and all materials of foundations to be made of very elastic india-rubber, and proportion them so that they will stand up and keep their shape: you will by those means diminish greatly the required thickness: THEN ADD 50 PER CENT (I. K. B. 1854)

Isambard Kingdom Brunel's involvement in the construction of over 1,200 miles of railway necessitated the building of hundreds of bridges, their structural form being dictated by the materials available, their structural potential and, as always, economic considerations. He used all three of the classical forms of construction which have remained

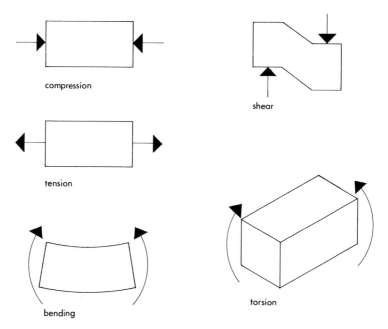

Fig 31 The five basic force actions

unchanged for thousands of years – the beam bridge, the arch bridge and the suspension bridge. The corresponding natural forms are the fallen branch or tree, a rock fall and the hanging vine or creeper. In order to obtain, at least in qualitative terms, an appreciation of the structural action of a particular form of bridge it is necessary to consider the force actions induced by the loads acting on it. Essentially the loads comprise the self or dead weight of the structure, imposed loadings due to vehicles, pedestrians and so on, and those due to the elements, in particular, wind action on suspension bridges. The physical effect of a force action is to produce deformation of the material fibres and the five force actions induced by the loadings – compression, tension, bending, shear and torsion – are shown diagrammatically in Fig 31. Compression and tension induce shortening and extension of the material fibres respectively and to quote Brunel: 'I have found that there is not a single substance we have to deal with, from cast iron to clay, which should not practically be treated strictly as a yielding elastic substance, and that the amount of compression or tension, as the case may be, is by no means to be neglected in practice any more than in theory.'

This is a statement of great significance as it shows Brunel's appreciation that compression and tension are the primary force actions, that all materials deform under load, and that the ground which supports the edifice should be considered as a structural material. Returning to the five force actions, it is apparent that bending, in the sense shown by the arrows, causes the top fibres to be compressed and the bottom fibres to be extended resulting in compression and tension. Shear is a measure of the tendency for racking or sliding of one part of the section relative to the adjacent part and torsion induces distortion or warping of the cross-section, both of which generate compression and tension. Thus it is possible to make a preliminary assessment of the structural potential of materials by considering their resistance to compression and tension:

MATERIAL	FORCE ACTION		UNIT WEIGHT	
	COMPRESSION	TENSION	kN/m^3	lb/ft^3
Timber	●	●	5	33
Brick	●		19	125
Masonry	●		20	130
Concrete	●		24	155
Wrought Iron	●	●	76	500
Cast Iron	●		76	500

Table 5 A comparison of the resistance of structural materials to force action and their unit weights

The materials available to Brunel were timber, bricks, masonry, concrete, cast iron and wrought iron. Of these six materials only two have a significant resistance to a tensile force action and their approximate unit weights should be noted.

Ten years after the birth of Marc Isambard Brunel, the first significant cast-iron bridge was constructed at Coalbrookdale in Shropshire. Although of a modest 100ft span, with proportions close to those of a masonry arch of the same period and detailing similar to that of high quality timber construction, its completion marked the end of the dominance of masonry, brick and timber as the construction materials for bridges. For many years the historical significance of Ironbridge was not appreciated, but fortunately it is now a central feature of the Ironbridge Gorge Museum complex which has won Europe's Museum of the Year award. However, cast iron, an alloy of iron and other elements which can be formed into intricate shapes when in a molten state, is only suitable for structures in which the dominant force action is compression and Brunel did not make extensive use of it in the superstructure of bridges: 'Cast iron girder bridges are always giving trouble ... The number I have is but few, because as I have before said, I dislike them.' (I. K. B. 1849)

Wrought iron is a purer form of iron which can easily be forged and shaped when hot, and in contrast to cast iron is more reliable in tension. It became commonly available in much greater quantities from the beginning of the nineteenth century and Brunel used it extensively for his major bridge constructions.

With regard to the use of masonry or brickwork he made the following observation: 'Bear in mind also that which is too often neglected and involves serious consequences, that masonry or brickwork has not half the strength which is generally calculated upon until the mortar is hard, and that you cannot keep centres or shores up too long.' From the above, it is apparent that Brunel had some sound practical advice to give on matters relating to construction materials. Further, he made extensive use of testing (see Technical Appendix) and, as we have seen in Chapter 1, was conversant with current developments in structural theory and the strength of materials, the quantitative origins of which date back to the Renaissance. The work of Galileo (1564–1642) formed the corner stone to the development of structural calculations, a brief description of which is included in the Technical Appendix as it forms a useful basis to a quantitative appreciation of structural mechanics.

It is apparent from Brunel's letters, calculation books and so on that he was familiar with the concepts of stress, strain and elasticity (see

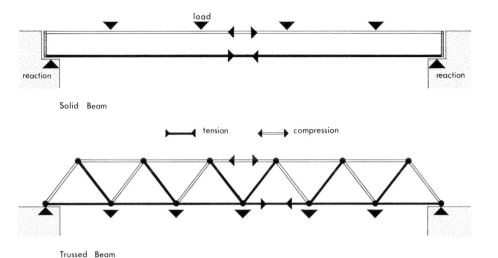

Solid Beam

tension ◀━━━▶ compression

Trussed Beam

Fig 32 The dominant force actions on a solid beam bridge and a trussed beam bridge

Technical Appendix) in both qualitative and quantitative terms and that stress levels used were, whenever practical, verified by experiment. It is in terms of application that he demonstrated the essence of engineering skill which will be illustrated by a description of a number of bridges which express his understanding of the structural potential of the materials currently available, and his imaginative use of one or more of the classical bridge forms – the beam, arch and suspension type. To recapitulate on these, the dominant force action on a solid beam bridge is bending under the action of the self-weight and vehicle loads (Fig 32). The total load is balanced by the reactions at the piers supporting the beams which are in turn subjected to a compressive action. An alternative to the solid beam is the trussed beam which consists of a triangulated assembly of compression and tension members. This results in increased structural depth but reduction in self-weight of the structure.

In the simple suspension bridge (Fig 33), the deck, a beam system, solid or triangulated, is supported by hangers or chains which are both tension members. The chains transmit compressive loads to the towers and at their extremities are anchored to massive foundations. Suspension bridges can be very flexible and deform excessively under the action of wind and vehicle loads unless the deck structure is stiff, that is deep. Shallow aerodynamically designed decks such as that used for the Severn Bridge are a recent development. The arch bridge may be considered as an inversion of the chain with the deck placed on top (Fig 34). The loads

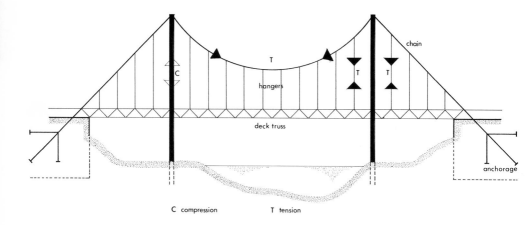

Fig 33 Suspension bridge

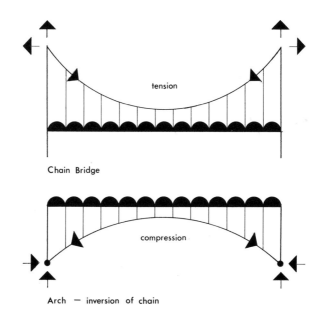

Chain Bridge

Arch — inversion of chain

Fig 34 Arch considered as an inversion of the chain

from the deck are transmitted via columns or walls to the arch rib which acts as a compression member. The arch compression or thrust C (see Fig 18 on page 52) is balanced by the foundation reaction R which has vertical and horizontal components. Thus the soil strata must resist outward and downward components H and V, the outward component H being large if the arch is shallow (see Technical Appendix). On soft ground this outward component induces a significant movement and thus to contain spreading of the arch rib, a tie is required (Fig 35). In brick or masonry construction, the gap between the arch rib and the deck is frequently connected by a series of spandrels with a gap between to reduce the self-weight of the structure.

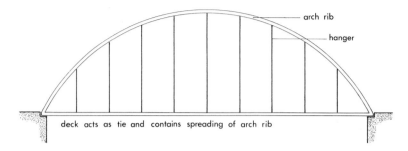

Fig 35 Tied arch

THE THAMES BRIDGE AT WINDSOR

In October 1849, the Great Western Railway completed a branch line from the main line at Slough to Windsor. This involved crossing the Thames by a wrought-iron arch/beam (bowstring) girder bridge of 203ft span and a maximum height of 23ft which was approached from the Slough side by means of a long, low timber viaduct (now brick). It crosses the Thames at a skew of 20 degrees and was originally designed for a double track, which has now been singled. This is Brunel's oldest surviving wrought-iron bridge and one of the outer trusses was subjected to a test load of 270 tons, uniformly distributed along its length. A cross-section through one of the girders is shown in Fig 36 and it can be seen that the arch rib is of cellular section in the form of a triangle. There are three arch ribs, and the central rib, which carries twice the load of the outer ribs, was made twice as strong by increasing the thickness of the plates. The outward thrust of the arch ribs is contained by the wrought-iron deck girders and thus the complete assembly could rest on simple abutments which were not required to sustain a horizontal load. The arch-beam system is stabilised by diagonal bracings in both the vertical

Fig 36 Windsor Bridge – a cross section through the main girder, turned through 90 degrees.

The brick approach viaducts to Brunel's wrought-iron bridge crossing the Thames at Windsor were originally of timber construction (*University of Surrey*)

Rocker connection between girders and abutment and timber flooring on the bridge at Windsor (*University of Surrey*)

and horizontal planes. Originally, the abutments supporting the girders consisted of six cast-iron cylinders (Fig 37), each girder being supported on two cylinders. The 6ft diameter cylinders were sunk by excavating the gravel from their interior by hand and placing weights on top to force them into the ground. The cast-iron cylinders have now been replaced by brick abutments and the rocker connection between the girder and abutment is shown in a photograph. The use of timber flooring can also be seen, and although these timbers are relatively new, Brunel frequently used timber flooring for bridge decks above which was placed ballast to support and distribute the loads from the track. Disadvantages of this detail are that the ballast added to the self-weight of the deck and also collected moisture which affected the durability of the timber decking.

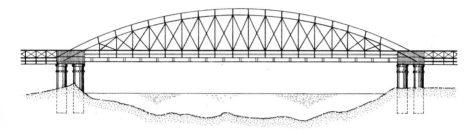

Fig 37 Arch ribs on Windsor Bridge were originally supported by cast-iron cylinders

Considering the structural form chosen by Brunel for this bridge, the cellular form chosen for the cellular arch rib is appropriate for a compression member, the restraint against buckling being provided by means of a vertical and horizontal bracing system. The deck is a deep flanged beam which can sustain the high bending moments induced by the moving loads. Starting with a solid rectangular section, the material can be more efficiently employed, whilst maintaining the same cross sectional area, by converting it to a deeper I or hollow section using thin plates. With the use of thin plates, the problem of buckling exists where compression occurs, and additional stiffening is often required. Due to the limited overall dimensions of plates available, they were riveted together by means of splice plates.

THE THAMES BRIDGE AT MAIDENHEAD
In Chapter 3 the general description of Brunel's elegant brick arch bridge crossing the Thames at Maidenhead and the controversy over its stability requires a more detailed technical appraisal (see also Technical

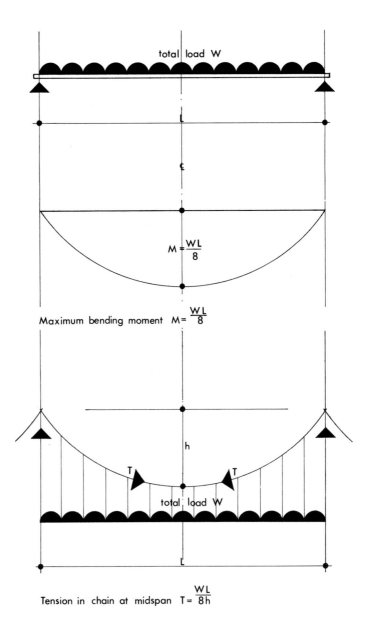

total load W

L

$$M = \frac{WL}{8}$$

Maximum bending moment $M = \frac{WL}{8}$

h

T   T

total load W

L

Tension in chain at midspan $T = \frac{WL}{8h}$

Fig 38 Parabolic variation in bending moment for a beam with total load W on a span L, maximum value $M = \dfrac{WL}{8}$

Fig 39 Tension in parabolic chain at centre $T = \dfrac{WL}{8h}$

Appendix). To minimize the height of the embankment and to obtain the required headroom, the main spans consisted of two shallow brick arches with spans of 128ft. There was a plentiful supply of bricks on this section of the line and, as we have seen, the arch is the appropriate structural form for long spans in brickwork as the dominant force action is compression. The major problems in arch construction are the need for centering, a temporary support system, generally timber, for the arch rib whilst it is being constructed and the need to contain outward movement of the arch caused by the horizontal component of the arch thrust. If we consider a simple beam with a total load W distributed uniformly over a span L, the maximum bending moment M at midspan is WL/8 and its variation over the span L is of parabolic form (see Fig 38). A light suspension chain subjected to the same loading W over a span L (Fig 39) will take up the same form as the bending moment diagram for the beam (see Technical Appendix) and the tension at the centre of the chain is WL/8h, where h represents the dip of the chain at midspan. The arch can be considered as an inversion of the chain and the horizontal thrust at the abutments is also WL/8h. Thus the horizontal thrust is related to the rise of the arch and increases as the ratio L/h increases. The above expressions apply to uniform loading which does not occur in practice and thus it is necessary to adjust the arch profile so that the thrust on the rib induces a uniform compression stress (Fig 40). This result can be achieved for permanent loading, that is the self-weight of the structure, but for moving loads, the line of action of the thrust will shift from the

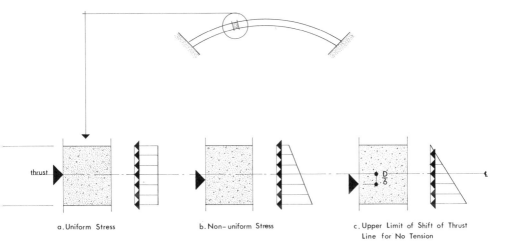

a. Uniform Stress   b. Non-uniform Stress   c. Upper Limit of Shift of Thrust
                                                        Line for No Tension

Fig 40 Stress distribution in an arch rib (a) uniform (b) non-uniform (c) upper limit of thrust line for no tension to occur in a rectangular section

central position on the rib causing non-uniform stress distribution. If this shift or eccentricity is large, the joints between the bricks will tend to open up as tension is induced in the section. Brunel remarked that 'in a very large arch, with a small rise, the line of pressure must be confined within very narrow limits'. For a rectangular section, the narrow limits are a distance of a sixth the depth of the section, if tension is to be avoided.

The arch profile chosen by Brunel for the Maidenhead Bridge is of semi-elliptical form (see Fig 41) with the rib depth increasing from

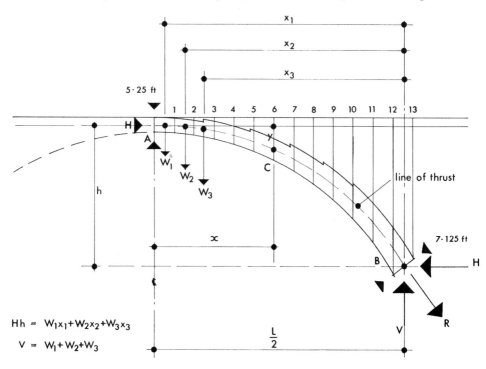

Fig 41 Arch profile chosen by Brunel for the Thames Bridge at Maidenhead

5.25ft at the centre line (crown) to 7.125ft at the springing or abutments. A simplified explanation of the analysis eventually adopted by Brunel is as follows. The half arch was divided into thirteen vertical sections and the weight of each section W_1, W_2, etc, estimated. Then if at any point C, a horizontal distance x from the arch centre line, the bending effect due to the thrust H times the distance y is made equal to the opposite bending effect due to the loads W_1, W_2, etc, to the left of C, then the bending effect is zero and the thrust induces a uniform compressive stress over the

depth of the rib. Thus at the arch springing, point B, we have the result
$H . h = W_1x_1 + W_2x_2 + W_3x_3 + \ldots$

Reference to Brunel's calculation books indicates that he developed a mathematical expression for the shape of the thrust line in terms of the distances x and y, having previously calculated the value of the horizontal thrust H. The horizontal thrust was equivalent to a stress of 10.25 tons/ft², a value well below the crushing strength of the brickwork. As we have seen in Chapter 3, the line of action of the thrust at the abutments of the main arches was diverted downwards by loading the smaller landward arches with concrete (Fig 42). Again this would achieve a more uniform distribution of stress on the foundations and Brunel writes: 'All forces should pass exactly through the points of greater resistance, or through the centres of any surface of resistance. Thus in anything resembling a column or strut, whether of iron, wood or masonry, take care that the surface of the base should be proportioned that the strain should pass through the centre of it.'

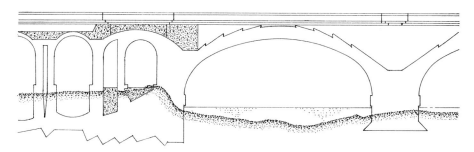

Fig 42 Landward arches loaded with concrete to divert the thrust from main arch

It can be concluded that Brunel had a sound understanding of the theories of arch design currently available and applied them in a highly professional manner to a number of brick arch bridges, with the one exception of the bridge over the River Parrett in Somerset (see Chapter 5). Maidenhead Bridge and Wharncliffe Viaduct are outstanding examples of his work and, as we have seen, the brickwork stresses are relatively low. This does not necessarily imply conservatism, as Brunel makes reference to the combined strength of bricks and mortar, and further the possibility of buckling cannot be overlooked, as a shallow arch rib should be treated as a slender compression member.

This brief technical appraisal of Maidenhead Bridge demonstrates, that contrary to popular opinion, Brunel did not use a 'back of an envelope' approach to the design of his major bridge structures. In his General Calculation Book there are detailed calculations along the lines described previously for a number of arch bridges which include the solution of complex equations of polynomial form. An inspection of Brunel's calculation and sketch books, complete with occasional cigar burns, gives an immediate impression of the many facets of his character – the sense of urgency suggested by the hurried scrawl, the superb draughtsmanship with meticulous attention to architectural detail, and the evolution of designs for complex structures built up from simple sketches. A number of major structures on the line between Paddington and Bristol were under construction simultaneously and this involved many hundreds of miles of travelling, lodging at scruffy inns and working until the early hours of the morning preparing calculations and reports. As the works proceeded, Brunel was able to make part of the journey by train with his britzska loaded on a special truck. The journey was completed by horse and foot. The number of structures involved necessitated miles of walking along the line to inspect and advise on the constructions and no doubt the presence of I. K. B. at a site greatly encouraged the work force.

CLIFTON SUSPENSION BRIDGE

Isambard Kingdom Brunel's first encounter with the design of suspension bridges was probably at the time he started work in his father's office in 1822 after returning from France. He was unable to enter the École Polytechnique as he was born outside France. His father was working on a project, commissioned by the French Government, for two suspension bridges to be constructed on Bourbon (now Reunion) Island in the Indian Ocean. The Mat River Bridge was a single span of 131ft with two roadways each of 9ft. The St Suzanne River Bridge was of similar dimensions but had two anchorages and a single tower at midspan. The bridges were fabricated in Sheffield and shipped out to Bourbon. At the time of construction of these bridges c. 1823, Thomas Telford's Menai Bridge was well on the way to completion with the world's largest span of 579ft. The standard of engineering of his father's designs for the bridges at Bourbon and Telford's impressive work at the Menai Straight were without doubt a source of inspiration to the then seventeen-year-old Isambard Kingdom Brunel, but a further six years elapsed before he was presented with the opportunity of preparing an independent design for a suspension bridge.

On 1 October 1829 an advertisement appeared in a number of newspapers inviting designs for the erection of an iron suspension bridge at Clifton Down over the Avon Gorge, to the west of the centre of Bristol. There was an enthusiastic response and over twenty entries were submitted, including four by Brunel who was at the time convalescing at Clifton. The four designs varied in span from 720ft to over 1,000ft. The Bridge Committee naturally sought the advice of Thomas Telford and he promptly rejected all the designs and produced his own scheme which consisted of massive Gothic style piers rising from the bottom of the gorge to support a three span suspended deck (see page 28). It is hard to believe that this design could follow the Menai Bridge but Telford was becoming cautious in his old age and was experiencing problems with oscillation of the deck and thus may have considered that any span greater than the Menai was too large. The central span in Telford's design was 400ft. Surprisingly the Bridge Committee approved his design but public opinion did not and Brunel commented: 'As the distance between the opposite rocks was considerably less than what had always been considered as the limits to which suspension bridges might be carried, the idea of going to the bottom of such a valley for the purpose of raising at great expense two intermediate supports hardly occured to me.'

Royal Assent was given to the Act for the Clifton Suspension Bridge in May 1830 with Telford's proposal deposited with the Parliamentary Bill. Fortunately, the committee decided to announce a second competition and co-opted the services of Davies Gilbert as judge. This was a wise decision as Gilbert was without doubt the best qualified person in Britain to make a technical assessment of the design of long span suspension bridges. He had previously advised Telford on the design of the Menai Bridge and recommended that the towers should be raised to reduce the force in the cables. His analysis of the bridge was based on the theory of the catenary, and in 1826 he published the first paper to provide a theoretical analysis for an unstiffened suspension bridge.

Of the twelve submissions, five were short listed – Telford, Brunel, Rendel, Hawks and Captain Brown, for further examination by Gilbert. He compared the designs on the basis of a crowd loading of 84lb per square foot of deck area and a safe tensile stress in the chains of 5.5 T/in^2. This safe tensile stress favoured Brunel's design and after much discussion he was formally appointed on 26 March 1831 to design and construct the bridge. This decision was influenced by the eloquence of Isambard Kingdom's arguments and his father's previous experience with the Bourbon bridges.

Brunel's calculations for the Clifton Bridge involved the use of simple statics, the tension in the chains being equivalent to the compression in the rib of an arch of similar inverted form. His calculation books demonstrate a knowledge of the geometry of the three known shapes, the parabola, simple catenary and catenary of uniform strength. He appreciated in qualitative terms the need for a stiff deck and had in mind to use additional cables under the deck to minimise undulation of the deck in a similar manner to those adopted by his father for the Bourbon bridges. In the period 1829 to 1850, Brunel's details for the design of the chains, deck and saddles at the top of the towers were continuously modified (Pugsley). At the early stages of the development of the design, he proposed the use of two chains, each with sixteen links. These were made up from rectangular plates with a lug at each end. The links were to be joined by pins passing through the overlapping lugs. Some years later he proposed the use of two sets of chains on each side of the deck and the hangers or suspension rods were to be connected to the chains by means of a triangular plate (Fig 43) to which were attached short rods which hang from each set of chains. The deck designs were essentially of timber construction, the stiffest of which consisted of 36in by 5in laminated longitudinal beams, above which were cross beams forming the deck and upper beams 12in by 5in in cross section (Fig 44).

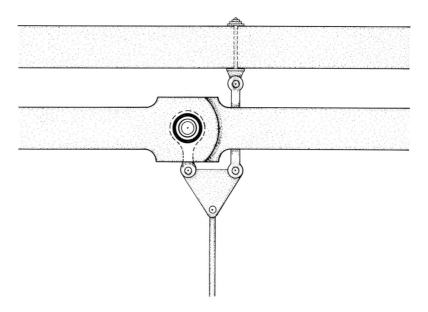

Link and Suspension Rod Arrangement

Fig 43 Brunel's two chain design with the hangers connected to the chains by a triangular plate

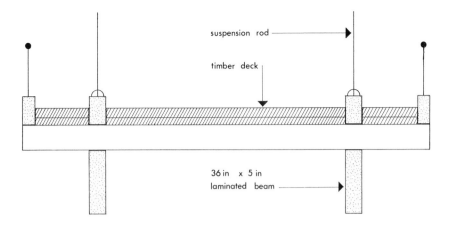

Deck Section for Clifton Suspension Bridge — circa 1850

Fig 44 One of Brunel's designs for a deck

Modifications to Brunel's design, introduced after his death in 1859, will be considered later.

Returning to the early stages of construction, which was always hampered by financial problems, the site of the bridge is about five miles from Avonmouth on a section of the Avon Gorge in the vicinity of Clifton Down and Leigh Woods. At this point the gorge is about 300ft deep and the land for the pier on the Clifton side became available in the summer of 1831. Work was commenced on the pier but progress was delayed due to a series of riots in Bristol. The foundation stone to the high Leigh abutment was laid in August 1836 and in that year a single iron bar, 1.5in diameter, was placed across the gorge, and Brunel travelled across in a basket swinging 230ft above water level. In the next twenty years progress was extremely slow and at the time of Brunel's death, the two piers had not been completed. Shortly after Brunel's death two leading members of the Institution of Civil Engineers, John Hawkshaw and William H. Barlow prepared a report on a project for completing the bridge. Brunel's Hungerford suspension footbridge, completed in 1843, was to be demolished and replaced by the present railway bridge connecting Charing Cross and Waterloo. The report proposed that the Hungerford chains could be used at Clifton. Subsequently a new bridge company was formed, with Hawkshaw and Barlow as engineers and naturally there were a number of modifications made to Brunel's final designs. The overall dimensions of the bridge, 702ft between the centre lines of the piers and the dip of the chains at 70ft were maintained, but the level of the road above high water was increased from 230ft to 245ft and the width of the deck was increased from 24ft to 30ft. Brunel's link

and suspension rod system (see Fig 43) was found to be unacceptable as it introduced bending in the upper chains, and to avoid this, three sets of chains were used on each side of the deck with hangers at 8ft intervals. The hangers were suspended from a 4.625in diameter pin which passed through the lugs at each end of the chains. The chains from the Hungerford bridge (central span 676ft) were not sufficient to provide the three sets of chains for the Clifton Bridge and a new set of chains were made for the uppermost layer. All the links for each chain were tested in tension to 10 tons per square inch which was about twice the estimated value with the full superimposed load on the deck. To reduce the total number of links required, the land chains on each side of the piers were steepened and thus they enter the anchorage tunnels closer to the piers. Hangers were not required for the land chains.

Brunel's laminated timber longitudinal girders (Fig 44) were replaced by wrought-iron plate girders constructed in sections 16ft long by 3ft deep on each side of the deck, separating the road from the footway. Construction of the bridge to the modified design was commenced in November 1862. Wire ropes were slung across the gorge to act as a temporary support to the chains. Work on the bottom layer of the eastern chains was commenced simultaneously from the Clifton and Leigh anchorages, and on completion the intermediate and top layers were constructed using the bottom layer for support. The chains were supported at the piers by means of saddles resting on roller bearings designed to accommodate the thermal movement. At the junction with the land, the chains are deflected by fixed saddles into tunnels, 60ft long, which were excavated at an angle of 45 degrees into the rock. Within the tunnels, the chains diverge (Fig 45) and are secured at the far end by means of massive cast-iron anchor plates.

Fixing the suspension rods (hangers) and longitudinal girders progressed simultaneously from both ends. The cross girders and diagonal bracing were then attached and the floor for the roadway consisted of 5in thick longitudinal and 2in thick transverse timber members. The parapet consists of iron lattice girders about 4ft high. It was indicated previously that Brunel intended to tie the bridge down from underneath but this was considered to be impractical as the vertical movement at the centre of the bridge can be about 6in. The total weight of the bridge between the piers including the chains, suspension rods, deck structure, roadway, and an allowance for traffic is in the order of

(*Previous page*) The external appearance of the Clifton Bridge has changed little since its opening in 1864 (*Frances Gibson Smith*)

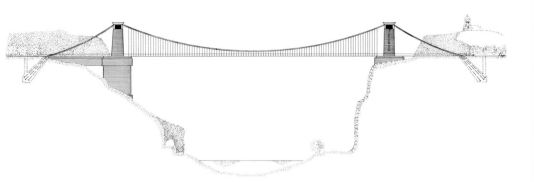

Fig 45 An illustration of Clifton Bridge showing the diversion of the chains in the tunnels

1,620 tons, and thus using the expression given in Fig 39 (see also Technical Appendix) for the tension in the chains at midspan then

$$\text{chain tension} = \frac{WL}{8h}$$

$$= \frac{1620 \times 702.25}{8 \times 70}$$

$$= 2{,}032 \text{ tons } (20{,}320\text{kN}).$$

At midspan there are thirty-two 7in by 1in links in the three sets of chains on each side of the bridge and thus the stress or intensity of load per unit area in the chains is

$$= \frac{2032}{2 \times 32 \times 7 \times .1}$$

$$= 4.54 \text{ tons/in}^2 \ (70.3\text{N/mm}^2).$$

At the piers the chain tension increases to about 2,190 tons and consequently the number of links is increased to thirty-five on each side of the bridge which results in a stress of 4.47 tons/in². This very approximate calculation meets the original requirement that the stress in the chains should not exceed 5.5 tons/in².

The bridge was opened on 9 December 1864, but prior to this it was tested by placing a dead load of 500 tons of stone over the road and footways. With a bridge of this type, subjected to continual movement in an exposed environment, careful maintenance is essential and the cradle for work underneath the deck is a well-known feature. The original timber decking has naturally been replaced as have the end cross girders which were placed too close to the masonry to be maintained properly

The land chains on the Leigh Woods bank with one set undergoing maintenance
(*Frances Gibson Smith*)

and thus suffered from corrosion. Corrosion has also occured in the anchorage tunnels, largely due to ingress of the water but this problem has now been resolved. Stress variations due to wind and the passage of up to 1,000 cars per hour necessitate frequent inspections for signs of fatigue cracking, but the original testing of the chains and bridge and the relatively low design stress levels have helped in this matter.

The external appearance of the bridge has changed little since its opening in 1864, and in its unique setting is a fitting tribute to its original designer but, as we have seen, it is not one hundred per cent Brunel. Brunel's last major bridge, the Royal Albert Bridge at Saltash, was designed for much more punishing railway loadings and is possibly his greatest technical achievement. It was opened in 1859 and is pure Brunel.

ROYAL ALBERT BRIDGE

As Engineer to the South Devon and Cornwall railway companies, Brunel was faced with the problem of spanning the River Tamar to form the rail link between Devon and Cornwall. The site chosen was Saltash, about three miles west of Plymouth where the river narrows to a width of about 1,100ft.

It was originally proposed to construct the bridge with one span of 255ft and six of 105ft, with trussed arch timber deck structures. Due to Admiralty requirements this design was modified to two spans of 300ft and two of 200ft again with timber deck structures. As warships were still under sail in the 1850s, a further requirement was that there should be 100ft of mast room below the soffit of the bridge at high water. The depth of the river reaches about 70ft at the centre, and thus the above proposals required the construction of a number of high piers to support the deck. Characteristically, Brunel contemplated the construction of a clear span of 1,000ft:

I have revised my calculation as to a span of 1,000ft, and find that even with the loads and limitations of strains which I adopt – namely a proper thickness of ballast, and a possible load of a train of engines without tenders, and a limitation under such load of 5 tons strain per square inch, that a span of 1,000 feet may be made in England of the very best workmanship, and sent out and erected for I should say safely £250,000, of course a single way, – another £250,000 ought, I should think, to cover the rest of the bridge. I should like to explain to you the mode I should propose for raising such a bridge, weighing 7,000 tons. (I. K. B. 1852)

About forty years previously, Marc Isambard Brunel designed a timber arched bridge of 800ft to cross the River Neva at St Petersburg and thus

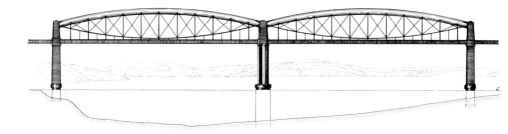

Fig 46 Royal Albert Bridge, Saltash, which was opened in 1859

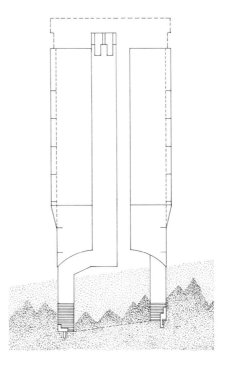

Fig 47 General arrangement of cylinder supporting central pier

both father and son demonstrated their confidence in designing large span bridges.

After a detailed examination of the rock strata underlying the mud on the river bed, and the pressure of severe financial stringency, Brunel's final design was for a single track bridge, of overall length 2,200ft with two main spans of 455ft and one central pier. There are seventeen approach spans ranging from 69.5ft to 93ft which are supported on masonry piers.

Work on the construction of the central pier foundation was commenced in early 1853. The central pier consisted of four octagonal

cast-iron columns, 100ft high, which were supported on a circular
column of solid masonry, 35ft diameter and 96ft deep, founded on the
rock below the mud. To achieve this, a wrought-iron cylinder was
constructed on the river bank, floated out into position and then, guided
by pontoons, lowered vertically into the river bed. The general
arrangement of the cylinder is shown in Fig 47 and using experience of
working underwater on the Thames Tunnel (Chapter 10), Brunel
provided a 4ft wide annular space around the bottom of the circum-
ference of the cylinder from which the water could be expelled by air
pressure. A ring of masonry was then built up and eventually a
reasonably watertight seal to the cylinder was achieved by constant
pumping. This meant that the central part could be completed in the
open. The inner plates of the annular ring, and subsequently the internal
access and air cylinders, were removed and the masonry built up to the
base level of the cast-iron piers. This work took three and a half years but

The two main spans of the Royal Albert Bridge, Saltash (*Frances Gibson
Smith*)

it must be remembered that the use of compressed air was a relatively new process (the shifts were limited to three hours to avoid 'the bends') and the lighting and tools were primitive.

Turning to the two main spans, they consist essentially, as indicated in Fig 48, of all three classical forms for bridge construction – the beam, arch and chain – the beam forming the deck structure which, by means of hangers, is connected to an arch/chain system, the outward thrust of the arch being balanced by the inward pull of the chains. About forty per cent of the chains used on this bridge were made over twenty years previously for use on Brunel's original design for the Clifton Bridge.

From experimental data, it was decided to adopt an elliptical wrought-iron tube for the arch, 16.75ft wide and 12.25ft deep. As we have seen, the tube is an appropriate form for a compression member, its elliptical form allowed smoother wind flow over its surface and further, the suspension chains could lie in a vertical plane and allow enough room for the passage of trains. As with the central pier, a major construction problem was involved. The complete assembly for each span was prefabricated on the Devon shore. With the customary Brunel thoroughness, each chain link was tested and the first complete span was loaded to in excess of 2.5 tons per foot run of its length in addition to its own weight of over 1,000 tons.

The floating out of the first arch/chain span took place on 1 September 1857, and the whole operation was directed by its designer from a platform in the centre of a truss. Even allowing for the fact that Brunel had the advantage of observing the floating and lifting of Robert Stephenson's Britannia Tubular Bridge girders some seven years previously, the event, witnessed by thousands, can only be described as heroic. By means of an elaborate system of signals, the truss was moved out into the river and then swung into position between the piers. Water was then admitted to the pontoons which lowered the truss onto the piers. The whole operation took two hours and was conducted in complete silence. This was followed by a vocal response which would now be reserved for kicking a ball in a net.

The truss was lifted in increments of 3ft at each end, followed by the masonry of the land pier, but it was not until July 1858 that it was lifted to its full height. Due to Brunel's illness, R. P. Brererton, his chief assistant, supervised the lifting of the second span and the bridge was officially opened on 3 May 1859 by Prince Albert. Brunel was not present

(*Opposite*) The central pier consists of four octagonal cast-iron columns (*Frances Gibson Smith*)

at the opening but shortly before his death he visited his last great bridge. His wish was to be remembered as an engineer and appropriately inscribed on the entrance portals to the Royal Albert Bridge is 'I. K. BRUNEL ENGINEER 1859'.

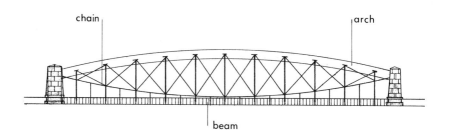

Royal Albert Bridge , Saltash   Primary Structural Elements

Fig 48 The main spans consist of the three classical forms for bridge construction

The general statement on design principles in Part 1 of the British Standard Specification for Bridges, published in July 1978, includes a clause which requires a useful life of 120 years with normal maintenance. The Royal Albert Bridge meets this requirement, and modifications since completion have not been excessive but have included the addition of horizontal lateral bracing half way up the hangers, stronger diagonal bracing and renewal of the approach span girders. It is now crossed by the 125mph High Speed Trains en route to Penzance but they are humbled to 15mph in deference to the designer.

7
THE RAILWAY VILLAGE AT SWINDON

Dear Sir,
I assure you that Mr Player was wrong in supposing that I
thought you purchased inferior coffee. I thought I said to him that I
was surprised you should buy such bad roasted corn. I did not
believe you had such a thing as coffee in the place: I am certain that
I never tasted any. I have long ceased to make complaints at
Swindon. I avoid taking anything there when I can help it.
Yours faithfully, I. K. B.

To many, Swindon is best known for Brunel's cutting remarks on the
standard of its station refreshment rooms, but of all the towns and cities
whose architecture has been influenced by the construction of railways
under Brunel's general direction, Swindon is unusual, in that a new town
was born as a result of the decision to establish a locomotive works on
section 7 (Faringdon Road to Hay Lane, see Chapter 3) of the Great
Western Railway between Paddington and Bristol. The old town of
Swindon, built on a hill, was for centuries the centre of a farming
community and is now bounded on the north by the railway and new
town and on the south by the M4 motorway. In the autumn of 1840,
acting on the advice of Brunel and Gooch, the directors decided to
construct the company's principal locomotive depot and repair shops in
the valley, a short distance to the north of the old town of Swindon where
work on section 7 of the line was nearing completion.

Other locations considered were Reading and Didcot but in a letter to
Brunel in September 1840, Gooch set out his views on the best site for the
principal railway establishment of the Great Western Railway. The three
primary reasons were, that it formed the junction with the Cheltenham
line, the nearby canal could be used as a source for supply of coal, coke
and water, and finally, Swindon was the point at which there was a
significant change in gradient. In the 77 miles between Paddington and
Swindon the average gradient was 1 in 1,320, but for the remaining 41
miles between Swindon and Bristol, steeper gradients were encountered.

Thus it was argued that a different class of engine would be required for the Bristol end of the line.

Regrettably, a number of the principal buildings associated with early development of the new town no longer exist – the passenger station and hotel, the engine shed and engine house. The approximate location of these buildings is indicated in Fig 49 and they are described in some detail by J. C. Bourne.

The passenger station, situated to the east of the junction of the Bristol and Cheltenham lines, consisted of two rectangular buildings 170ft by 37ft on each side of four lines of railway which were roofed over. Each building was constructed on three levels – a basement accommodating kitchens and service rooms, the main floor at platform level offered refreshment facilities for first and second class passengers only, and the top floor was used as a hotel with bedrooms in the northern building and coffee and sitting rooms in the southern building. The two sections of the hotel were connected by a covered way.

The engine shed and engine house were located a short distance to the west of the passenger station in the fork between the Bristol and Cheltenham lines. The former was of rectangular form 490ft long by 72ft wide, open at both ends and its wooden roof contained louvres to allow the escape of steam. It was in this building that engines were prepared for work. The engine house, used for carrying out light repairs, was erected at right angles to the engine shed and J. C. Bourne's lithograph gives a clear impression of the structure of this impressive building. Again of rectangular form, 290ft by 140ft, it consisted of three bays separated by two rows of timber columns. In the 50ft wide central

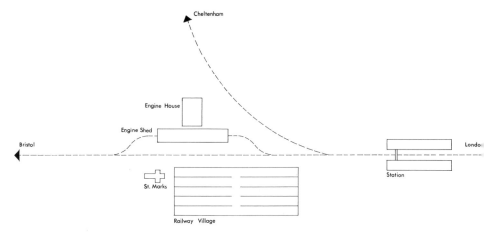

Fig 49 The approximate location of the main buildings at Swindon

The engine house at Swindon in a lithograph by J. C. Bourne

bay, a traverser was used to transfer the engines to any one of the stalls in the outer bays. The roof structure consisted of substantial timber rafters acting in conjunction with slender wrought-iron tension members. This was an excellent example of Brunel's use of timber and wrought iron to form a large span roof, purely functional and free of the usual embellishments in various architectural styles. Further buildings constructed in the vicinity of the engine house were the major repair shops and an erecting house.

Roughly opposite to these principal buildings and to the south of the railway line, it was decided to build cottages and other facilities for the company's employees, and as a result of imaginative public-spirited action by the Borough of Swindon (now Thamesdown Borough Council), the cottages are being modernised and restored. Apart from a few pioneers such as Robert Owen (1771–1858) at New Lanark, and Sir Titus Salt at Saltaire, the nineteenth century is not renowned for examples of projects in which serious attempts were made to improve the welfare or travelling conditions of the poorer classes associated with industrial development. The Great Western Railway in its early days was reluctant to carry third-class passengers at all and for a number of years

A street in the railway village at Swindon (*University of Surrey*)

they were accommodated in open trucks. Subsequent covered accommodation was primitive and no windows were provided. In contrast to this, the company had purchased sufficient land for the construction of three hundred cottages which were built by Messrs J. & C. Rigby of London. They were also responsible for the construction of the refreshment rooms, hotel and locomotive buildings. In order to avoid the building costs, an agreement was made between the builders and the railway company, that the builders should carry out the work at their own expense, but in return would obtain rents from the cottages and receipts from the hotel and refreshment rooms. The lease on the buildings was charged at the rent of one penny per year. With a contractor's eye for a quick profit, J. & C. Rigby immediately sublet the refreshment business and in 1848 sold the remaining 92 years of the 99 year lease for £20,000.

In spite of the 'wheeling and dealing' over the premises at Swindon, the outcome was a railway village which is yet another example of Brunel's imagination and further, his desire to provide reasonable living conditions for the work force. Brunel worked in conjunction with the architect Matthew Digby Wyatt on the design of the cottages, and it was

An alley in the railway village at Swindon (*University of Surrey*)

(*Opposite above*) The Glue Pot in Swindon railway village, formerly Thomas's Beer House (*University of Surrey*)

(*Opposite below*) The entrance to the Mechanics' Institute in Swindon village (*University of Surrey*)

Wyatt who later assisted him on the design of Paddington Station. The general layout of the cottages and associated buildings is shown in Fig 50. The parallel rows of cottages are separated by wide streets and a number of different styles were adopted. Extensive use was made of Bath limestone which had become available during the construction of Box Tunnel. Design features such as recessed windows with an exterior splay, a slate roof and diamond-shaped chimneys, were also used on Brunel's minor stations such as Pangbourne and Culham (see Chapter 8). Another interesting design feature sometimes used was the doors to two cottages being recessed and set at about forty-five degrees to the outside walls. At the end of the rows of cottages facing Exeter Street and Taunton Street, the central alley passes through a house at each end with a Gothic-style exit. Although drinking by the 'lower orders' was not supposed to be encouraged, three pubs were built, the Bakers Arms, the Cricketers and Thomas's Beer House, now called the Glue Pot.

In the central portion of the village between the two blocks of cottages, further amenities were provided, the Mechanics' Institute, a market and

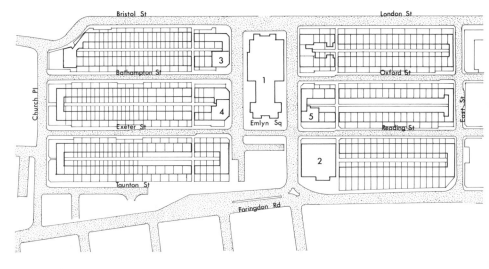

1 Mechanics Institute   2 Great Western Railway Museum   3 Bakers Arms p.h.   4 Cricketers Arms p.h.   5 Glue Pot p.h.

Fig 50 Swindon railway village

St Marks Church, Swindon (*University of Surrey*)

a hospital. The Mechanics' Institute, built in 1855 and enlarged in 1892, provided evening classes in mathematics, engineering, science and literary and domestic subjects. In 1854, the company built a one hundred bedroom hostel which later became the Swindon Wesleyan Chapel and subsequently the Great Western Railway Museum. Amongst the exhibits concerned with the works of Brunel and Gooch, the museum houses a full-size replica of the *North Star*, one of the best known of the broad gauge engines. It hauled the first passenger train to run from Paddington to Maidenhead (see Chapter 9). To cater for the spiritual needs of the community St Marks Church was completed in 1845 and it is located at the western end of the village. It was designed by Gilbert Scott, who was later knighted for his services to architecture.

In 1965, there was considerable speculation with regard to the future of the railway village and talk of it being replaced by a shop and office complex. Fortunately, the freehold on the majority of railway village cottages was purchased by Swindon Borough Council from British Rail for about £120,000. This worked out at about £300 per cottage and in 1969, modernisation was commenced. The photographs demonstrate that the essential character and appearance of the village has been preserved. The renovation included replacing the original timber ground floor by a concrete slab, treating the walls for rising damp, space heating and roof insulation. As far as possible the original roof slates were re-used and supplemented with asbestos slates. The project has justifiably received a number of awards.

A 'third' town has now been developed between the railway village and the old town and incorporates the Brunel Shopping Centre with spacious arcades built in the Victorian style. In the centre of the shopping complex is the inevitable statue of the man who conceived the new town of Swindon some 120 years previously.

8
STATIONS

I am going to design, in a great hurry, and I believe to build, a Station after my own fancy; that is with engineering roofs etc. It is at Paddington, in a cutting, and admitting of no exterior, all interior and roofed in. . . . Now such a thing will be entirely metal
(I. K. B. in a letter to Digby Wyatt 1851)

Isambard Kingdom Brunel's involvement in the construction of over 1,200 miles of railway necessitated the construction of many railway stations. It was unusual for a civil engineer to be responsible for the design of secondary buildings such as stations and engine sheds, and Brunel's concern with track layouts, station buildings and even lamp posts is characteristic of his thoroughness. On the line between Paddington and Bristol twenty-four stations were constructed on the nine sections completed between 1838 and 1841. This excludes the stations at Didcot (1844) and Swindon (1842), which were opened shortly after the operation of through trains between the London and Bristol termini. It is fairly certain that Brunel personally designed these stations and it is possible to classify them into a number of different types (Vaughan, Bourne) – principal stations, one-sided stations, and second-class or minor stations. Vaughan has categorised the minor stations on the Paddington to Bristol line and elsewhere into four styles, related to the design of the roof and awnings. Typical examples of these styles are the stations at Pangbourne and Culham which will be described later. The 'Brunelian' influence can be seen in the design of numerous stations on secondary routes such as the Wilts, Somerset & Weymouth Railway and the Reading to Basingstoke line. Table 6 classifies the stations between Paddington and Bristol into the three types.

Apart from Paddington (new station) and Bristol Temple Meads, little remains of a unique contribution to station architecture, conceived with the overall aim of providing an environment to fulfil passenger and business needs. This was coupled with the practical requirement of standardised designs and the appropriate utilisation of the construction

PRINCIPAL STATIONS	ONE-SIDED STATIONS	MINOR STATIONS
Paddington	Slough	Ealing
Didcot	Reading	Hanwell
Swindon		Southall
Chippenham		West Drayton
Bath		Taplow
Bristol Temple Meads		Twyford
		Pangbourne
		Goring
		Wallingford (Moulsford)
		Steventon
		Faringdon
		Shrivenham
		Hay Lane (temporary)
		Wootton Bassett
		Corsham
		Box
		Saltford
		Keynsham

Table 6 Brunel designed stations between Paddington and Bristol

materials available, that is, brick, flint, limestone, timber, and metal. For a more detailed description of Brunel's contribution to station architecture, it is convenient to follow the route of the Great Western Railway from Paddington to Bristol. The new station at Paddington (1854) is a superb example of London station architecture, but even this was surpassed by W. H. Barlow's (1812–1902) masterpiece, the St Pancras train shed (1865–8). Barlow was also involved in the completion of the Clifton Suspension Bridge (see Chapter 6).

Principal Stations

PADDINGTON
A few months after the opening of section 1 of the Great Western Railway in 1838 (Chapter 2), Paddington's first station was completed (Clinker). It was located immediately to the west of Bishops Road, which in the vicinity of the station was supported by a series of arches. The sites of the old and new stations are shown in Fig 51. The arches supporting Bishops Road formed the entrance to the old station which consisted of two departure and two arrival platforms roofed over by triangulated

arched trusses. The old station acted as a terminus for sixteen years, but for most of its life was inadequate for the needs of passengers. Thus it was decided to construct a new station to the east of Bishops Road within the L bounded by the junction of Eastbourne Terrace and Praed Street. In the design of Paddington new station, Brunel was assisted by the architect Digby Wyatt who had also worked on the railway village at Swindon (Chapter 7). The ten lines of broad gauge track and four platforms were covered by three wrought-iron arch ribs, the departure side, adjacent to Eastbourne Terrace, being opened in January 1854 and the arrival side, five months later.

The construction of the roof was strongly influenced by the work of Joseph Paxton, whose design of the metal and glass structure for the Great Exhibition of 1851 (the Crystal Palace) was, along with Brunel's prefabricated Renkioi hospital, one of the first examples of the concept of industrialisation of the building process. Brunel and Digby Wyatt were both involved with various aspects of the Great Exhibition and were naturally impressed by one of the most innovative designs of the nineteenth century. Paxton devised a means of semi-mechanised fixing of the glass roof lights for the Great Exhibition structure, and Paddington Station roof was partially covered with Paxton glass roof lights. In his letter to Digby Wyatt (January 1851) concerning the construction of Paddington Station, Brunel states that he has neither the

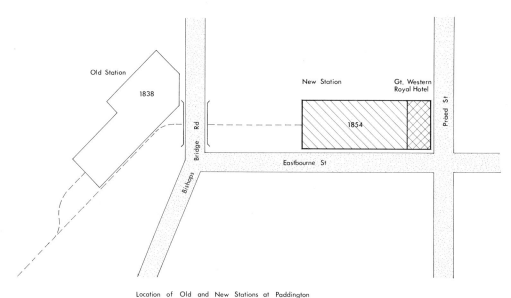

Location of Old and New Stations at Paddington

Fig 51

The first arch which spans platforms 1 and 2 at Paddington. Tie rods added *c*.1915 (*University of Surrey*)

time nor the knowledge for detail of ornamentation and suggests an evening meeting: 'Do not let your work for the Exhibition prevent you. You are an industrious man, and night work will suit me best . . . I shall expect you at $9\frac{1}{2}$ this evening.' This characteristic letter expresses Brunel's haste and views of the role of the architect in station design.

Paddington Station, as completed in 1854, was 700ft long and the three arch ribs were supported by two internal rows of cylindrical columns to give spans of 68ft, 102ft and 68ft respectively. The basic layout of the station remained the same for over fifty years, an additional fourth bay being added between 1913 and 1915, and the original cast-iron circular columns were replaced by hexagonal columns about ten years later.

The first arch which now spans departure platforms 1 and 2 has changed little since its construction in 1854. One modification has been the addition of tie rods (*c*. 1915) which were rather clumsily connected to the arch ribs. The apparent sag of the tie rods indicates they are carrying

little of the horizontal thrust from the arch ribs. The roof is crossed at two points by trancepts 50ft wide, opposite one of which is the elaborately detailed directors' balcony. The combination of Brunel's engineering and Digby Wyatt's ornamentation is vividly expressed in the view taken from the footbridge crossing the platforms at the western end of the station. The wrought-iron arch ribs at 10ft spacing spring from double diagonally braced trusses which span 30ft between the hexagonal columns. The decorative work in cast iron is bolted onto the webs of the arch ribs.

The High Speed Train gliding smoothly into Paddington Station a little over one hour after its departure from Bristol Parkway is a typical example of British Rail's programme of providing passenger comfort and speed in the tradition conceived by Brunel. Unfortunately the haste of travel in the 1980s allows little time to spare even a few minutes to admire the station roof which demonstrates so admirably his unique talents.

The Great Western (Royal) Hotel is connected directly with the station concourse and was officially opened in June 1854. The hotel and the terminus were the starting point of Brunel's concept of a complete transportation system between London and New York offering perfection in travelling, but for the 'higher orders' only. It was designed by P. C. Hardwick in the style of a French château and with over a hundred bedrooms and sitting rooms, public rooms, lounges and a restaurant, the hotel was, at the time of completion, Britain's largest. The cost of the hotel including all the finishes and fittings was around £60,000, and the first chairman of the board of directors was Isambard Kingdom Brunel. The hotel was managed independently of the railway company, and with Brunel as chairman a high standard of service could be taken for granted. The hotel's brochure (Vaughan) emphasised that 'passengers by the trains can pass between the platforms and the hotel at once, without trouble or expense, and proper persons will be always in attendance to receive and carry the luggage to or from the trains'. It was possible to obtain a small bedroom on the fourth floor for as little as 1s (5p) per day, and draught port was 2s 6d (12½p) per pint.

Since its opening in 1854, the hotel's tariff has changed dramatically

(*Opposite above*) The detail of the connection between the tie rods and arch rib, platform 1 (*University of Surrey*)

(*Opposite below*) The directors' balcony over platform 1 at Paddington Station (*University of Surrey*)

The Great Western Royal Hotel at Paddington with the château towers
(*Frances Gibson Smith*)

and the cost of a single room is now (1979) over three hundred and fifty times that of 1854. It has been modernised on a number of occasions and lost much of its decoration, but the château towers still form an impressive frontage to the station. In the autumn of 1979 the new Brunel Bar and Restaurant were opened and designed to a Victorian theme. The restaurant offers specialities such as 'scallops of veal Brunel' as well as a range of traditional dishes. The hotel's publicity states that Brunel would have been proud of the new bar which is not in any way 'toffee nosed'. The author is inclined to think that the availability of real ale in half-pint glasses only would not have met with the approval of the hotel's first chairman, as in 1854, it was apparently gentlemanly to order port by the pint. To stay at the Great Western is a pleasant experience and contrasts with the computerised environment of many of London's more recent hotels.

On Thursday 1 March 1979, the 125th anniversary was celebrated by a number of events. A special train hauled by the preserved King class

4–6–0 *King George V* was scheduled to run to Didcot and back, but regrettably it was unable to manage the return journey due to an overheated axlebox on one of the driving wheels. The Swindon brewers, J. Arkell & Sons advertised a special anniversary brew, and there was an exhibition and photographic display. An official history by C. R. Clinker was published by British Rail Western Region.

DIDCOT

Travelling westwards from Paddington, the first principal station with an overall roof was opened at Didcot on 12 June 1844, some three years after the completion of the line to Bristol. It was of timber construction and consisted of five platforms and four lines (Macdermot), and a general impression is given in Fig 52. It was destroyed by fire in 1885 and subsequent constructions bear no resemblance to the original, which in itself was not a typical Brunel design.

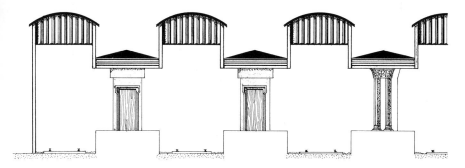

Fig 52 An impression of Didcot Station with an overall roof, *c*.1844

SWINDON

Swindon Station has been described in Chapter 7. It was built without an overall roof, but the platforms were protected from the weather. The station now consists of a single island platform approached via a tunnel under the railway.

CHIPPENHAM

Chippenham and Swindon were the two principal stations between Paddington and Bristol which were not built with overall roofs. An illustration of Chippenham Station *c*. 1841 (Vaughan) has a number of details which are similar to those on Mortimer Station. The forebuilding but not the platform structure at Chippenham may, in part, date from the opening of the line.

BATH

The architecture of the original stations at Bath and Bristol Temple Meads had a distinctly medieval style, and it is unfortunate that the overall roof to Bath Station was removed in the last decade of the nineteenth century. In the design of both station roofs it can be argued that Brunel disguised their true structural form with the use of hammer-beam detail. The structural action of a hammer-beam roof is a matter of some controversy, but of the possibilities, one is that the hammer-beam (Fig 53) acts as a cantilever supporting the load from the hammer post.

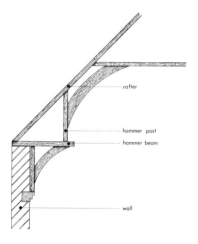

Fig 53 Hammer beam roof

The inner face of the wall acts as a fulcrum and the cantilever is prevented from rotating by the thrust of the rafter at its outer end. J. C. Bourne's lithograph gives a good impression of the interior of Bath Station and he describes more correctly the structural action of the roof as follows:

The peculiarity of this station lies in its roof, which is 60ft span without either buttress or tie of any description, either vertical or horizontal; the ribs or principals being in fact the long arms or jibs of a series of cranes, of which the side columns represent the upright parts. These jibs meet in the middle, and are still further steadied by the cross-diagonal planking of the roof. Nothing can be more simple, or better suited to the general appearance of the station.

(*Opposite above*) Chippenham Station, the forebuilding (*Frances Gibson Smith*)
(*Opposite below*) Bath Station, in a lithograph by J. C. Bourne

Bourne's description of the roof at Bristol Temple Meads suggests a similar structural action.

BRISTOL TEMPLE MEADS

In order to raise the railway to an adequate level to pass over the 'floating harbour' (Fig 54), Brunel constructed the Bristol terminus of the Great Western Railway on a series of arches about 15ft high. The first station building was completed on 31 August 1840 and for a number of years it was also used by the Bristol & Exeter Railway. In 1845 a separate terminus was built at right angles to the first station building and some years later, a joint station was built by the Great Western, Midland and Bristol & Exeter companies. It is proposed to describe the first station building only which extends down to Victoria Street. The interior of the station now suffers the indignity of being used as a car park.

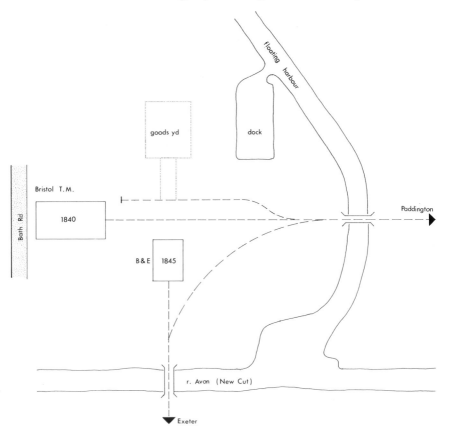

Fig 54 Location of Bristol Temple Meads Station (1840) in relation to the Floating Harbour, River Avon and B & E station (1845)

Bristol Temple Meads Station, completed in August 1840 (*Frances Gibson Smith*)

Although in need of cleaning up and some minor repairs, the roof remains as Bourne described it in the 1840s when it covered five broad gauge tracks:

The roof is peculiarly well suited to the purposes to which it is applied; and as it covers a clear span of 74ft, without the aid of either cross tie or abutment, a particular notice of it will not be out of place here. It is composed of a series of 44 ribs, 22 on each side, and placed 10ft apart each of which is constructed somewhat like the jib of a crane, that is to say, of a long arm, projecting far and rising high into the air, and a short arm or tail, which in heavy cranes is either tied down or loaded. In the present case the iron columns which divide the central space from the aisles are the fulcra or crane posts upon which the arms rest [See Technical Appendix.] The long arm or jib extends to the centre and ridge of the roof, and there meets its fellow from the opposite side, while the short arm or tail is carried backward to form the roof of the aisle, to the outer wall of which it is held down by a strong vertical tie passing some way down. The whole is then planked over diagonally and is intended to be fitted up and decorated to suit the rest of the building.

The interior of Bristol Temple Meads Station (*Frances Gibson Smith*)

This is a plausible description of the structural action of the roof and, as with Bath Station, the hammer-beam detail is merely decorative. An intriguing feature of Brunel's design for the roof of Bristol Temple Meads Station is the question of whether it can be considered as a genuine timber structure. Behind a missing pendant may be seen an iron column. Further, it appears that the principal roof members were plated on both sides (Booth, Totterdill). Perhaps it would be more accurate to describe the roof as a composite construction of metal and timber, but this does not detract from the boldness of its design, the appearance of which and certainly its structural integrity would be improved by the omission of the hammer-beam detail.

Fortunately there are plans by British Rail and conservationists to restore the grade 1 listed Temple Meads Station. It will require an estimated £1.5m to restore it to its former glory and presents an opportunity for carrying out a detailed inspection of the roof members and the hammer-beam detail. In its present state it is the earliest surviving major main line terminal station from the railway mania era still in its original state.

One-Sided Stations

The route of the Great Western Railway (Chapter 3) was such that the towns of Slough and Reading were situated mainly to the south of the line, and Brunel adopted his one-sided design which he considered to be a convenient arrangement for passengers. The up and down platforms were built on the south side of the line and were in effect separate stations. It is apparent from the simplified layout shown in Fig 55 that the up stopping trains crossed the down main line twice and also crossed the down platform loop. Complete platform cover was provided at both Reading and Slough but increased traffic caused excessive delays and they were replaced before the end of the nineteenth century.

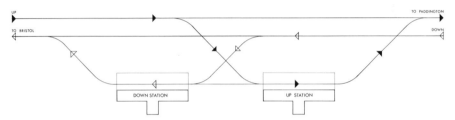

Fig 55 Simplified layout of a one-sided station

Minor Stations

CULHAM AND PANGBOURNE

Quadrupling of the line and more recently the withdrawal of stopping train facilities led to the demolition of all the original minor stations between Paddington and Bristol. Bourne selected Pangbourne as an example of a second-class or minor station:

The station, as is usual with those of this class, is composed of a house, placed on the side of the railway most accessible to the public, and which here happens to be the north, a covered platform being placed on the opposite side. The station house is constructed of flint rubble, neatly pointed, with quoins, door and window cases, and dressings of red brick. The general style is Elizabethan. It consists of one storey only, and is divided into a booking office with a bow looking upon the railway, and two waiting rooms. The eaves of the building are produced so as to form a complete covering all round the house, and extending over the platform towards the railway.

(*Above*) Pangbourne Station in a lithograph by J. C. Bourne

(*Opposite*) Culham Station on the Didcot to Oxford line (*Frances Gibson Smith*)

It is fortunate that the line between Didcot and Oxford has never received the traffic to necessitate quadrupling the line, otherwise probably the last surviving example of the station type illustrated in Bourne's lithograph of Pangbourne Station would have been demolished. Culham Station, a few miles north of Didcot, is similar to Pangbourne apart from its cast-iron brackets below the awning support beams. The covered building on the platform opposite the station house, as shown in Bourne's illustration of Pangbourne Station, has been replaced at Culham by what can be described as a prefabricated hut with no awning.

MORTIMER

Mortimer Station on the Berks & Hants line incorporates a number of Brunelian features which fall into one of the styles related to the design of the roof and awnings. In contrast to Culham Station, the roof is low pitched but the awning support beams are maintained without the cast-iron brackets. Of particular interest is the door opening at the gable end of the station house with the circular arch over. This detail and that of the timber door, are identical to that shown in an illustration of Chippenham Station *c.* 1841 (Vaughan). Mortimer Station was opened in 1848.

(*Above*) Mortimer Station on the Berks & Hants line (*Frances Gibson Smith*)

(*Opposite above*) Frome Station (*Frances Gibson Smith*)
(*Opposite below*) Frome goods shed (*Frances Gibson Smith*)

FROME

Frome Station on the Wilts, Somerset & Weymouth Railway was opened in 1850 and was designed by one of Brunel's thirty or more assistants, J. B. Hannaford. The station has an overall roof and is a unique surviving example of this type of construction. The use of timber rafters and wrought-iron ties should be noted. The column supporting the gable structure was added after 1973, presumably due to deterioration of the exposed timber. Of equal interest is the goods shed which has a more cumbersome timber roof.

9
GREAT WESTERN LOCOMOTIVES
(1837–59)

Lastly, let me call your attention to the appearance – we have a splendid engine of Stephenson's, it would be a beautiful ornament in the most elegant drawing room and we have another of Quaker-like simplicity carried even to shabbyness but very possibly as good an engine, but the difference in the care bestowed by the engine man, the favour in which it is held by others, even oneself, not to mention the public, is striking. A plain young lady however amiable is apt to be neglected. Now your engine is capable of being made very handsome, and it ought to be so. (I. K. B. 1838)

The early developments of steam locomotion in the pre-locomotive era were briefly described in Chapter 2, and during the early years of operating the train service from Paddington to Bristol, Brunel was very much indebted to the expertise in locomotive design which had been developed in Britain over the previous three decades, and in particular the contribution of the Stephensons.

The first known attempt to use steam as a source of power dates back to the first century AD, but it was not until the eighteenth century that the steam engine was established as a reliable source of power (Singer *et al*) by Thomas Newcomen (1663–1729). Newcomen developed a stationary steam engine which operated at atmospheric pressure (see atmospheric railway, Chapter 4). During the first quarter of the eighteenth century, the Newcomen engine came into general use in England and elsewhere for pumping water in mines. In the latter half of the century, James Watt (1736–1819) in partnership with Matthew Boulton was able to improve the efficiency and mechanical design of the steam engine. A major contribution to these improvements can be ascribed to John Wilkinson's (1728–1808) patent for machinery capable of boring cylinders to fine limits, since the efficiency of Newcomen engines had been impaired by ill-fitting pistons. However, the Watt engine still operated at low pressure and was of course stationary, but in 1784 he obtained a patent for a locomotive steam engine. The Watt locomotive was never developed, and it was Richard Trevithick (1771–1833) who built the first steam locomotive to run on rails. In 1803, he constructed the single cylinder Coalbrookdale locomotive which ran on cast-iron plate rails and the boiler pressure was 50lb per square inch roughly 3.5 times that of the atmosphere. The use of high pressure was a major innovation in the design of steam engines. Nine years later, John Blenkinsop (1783–1831) was the first to design a locomotive with two cylinders, but its more widely known feature was a rack wheel drive on to cogs on the side of the rail.

In 1814 George Stephenson constructed the 0–4–0 Killingworth locomotive whose performance has been described (see Chapter 2) but it was not until the Rainhill trials of 1829 that the locomotive was fully established as a reliable alternative to the horse. On a two-mile stretch of the nearly completed Liverpool & Manchester Railway at Rainhill, George and Robert Stephenson's 0–2–2 *Rocket* satisfied the conditions of a competition for the best form of engine, and its two serious rivals the *Novelty* and *Sans Pareil* failed to fulfil the requirements. The success of the *Rocket* was due to the introduction of a multitubular boiler containing twenty-five copper tubes of 3in diameter.

Before describing the significance of this, there is a further Stephenson engine to be considered. The 2–2–0 *Planet*, built in 1830 for the Liverpool & Manchester Railway, incorporated the essential features of the 'modern' steam engine from which there has been little significant development. These features were:

1 Horizontal cylinders
2 A multitubular boiler containing 129 tubes of $1\frac{5}{8}$in diameter (G. Drysdale Dempsey) which dramatically increased the boiler heating surface and the production of steam
3 The use of the blast of the exhaust steam to create a draught for the fire

The boiler pressure was 50lb per square inch and the external wooden frame was strengthened by metal plates. These features were adopted in subsequent locomotive designs for the Great Western Railway. In the mid 1830s, the Stephensons were incorporating valve gearing to control the admission and release of steam from the cylinders and the firm of R. Stephenson & Co, Newcastle, gained an international reputation as locomotive builders.

In 1836, Brunel laid down some specifications (Macdermot) which the locomotive builders were requested to comply with, one of which was that his 'standard' locomotive velocity of 30mph should be obtained at a piston speed of 280ft per minute. It is established in the Technical Appendix that the relationship between the piston speed V_P ft/min and locomotive speed V_T mph is given by the following expression:

$$V_p = 56.05 \frac{V_T S}{D}$$

$$\text{thus } D = 56.05 \frac{V_T S}{V_p}$$

where S = stroke of piston in feet
D = driving wheel diameter in feet

The above expression can be used to demonstrate why the locomotive builders were forced to adopt such large driving wheel diameters. One of the earlier locomotives delivered to the Great Western Railway was 2–2–2 *Ajax* built by Mather, Dixon & Co in 1838 (P. J. T. Reed, Macdermot), see Fig 56. The diameter of the driving wheels was 10ft and the piston stroke 20in. Using the above equation, the following value for the wheel diameter is obtained using a piston speed of 280ft per minute and an engine speed of 30mph specified by Brunel:

$$\text{Driving wheel diameter } D = \frac{56.05 \times 30 \times 1.67}{280}$$

$$= 10.03\text{ft}$$

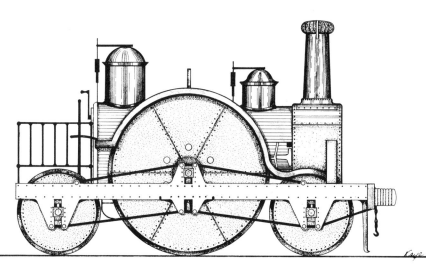

Fig 56 A GWR 'freak' engine 2-2-2 *Ajax* (1838) with 10ft driving wheels

The illustration of *Ajax* suggests why the first series of unclassified engines delivered to the Great Western Railway were described as freaks. The *Ajax* had a working life of less than two years – the general performance of Great Western engines between 1837 and 1840, with the exception of six designed by R. Stephenson & Co, was, at best, poor.

It was fortunate for Brunel that Daniel Gooch (1816–89) applied in 1837 for the post of locomotive superintendent of the Great Western Railway. Gooch's first task was to maintain a service on the first section of the line which was opened in 1838 between Paddington and Maidenhead. Previous experience with R. Stephenson & Co provided a sound background to Gooch's career as locomotive superintendent, and in November 1837 the 2–2–2 *North Star* was delivered by barge to Maidenhead. In May 1838 it worked the first passenger train and became

An illustration of the 2-2-2 *North Star*, delivered to the GWR in November 1837 (*P. Holt*)

established as the most reliable engine on the line. The *North Star* was built by R. Stephenson & Co, and between 1839 and 1842, eleven further Stars were brought into use. Brunel's praise of the *North Star* quoted at the beginning of this chapter was fully justified. It is probably the best known of the Great Western broad gauge engines and has a long history. It was one of two engines built for the New Orleans Railway of the USA to 5ft 6in gauge but it was never delivered. It was purchased by the Great Western Railway and the gauge and driving wheel diameter altered. After twenty-seven years' service, it was rebuilt in 1854 and ceased work in 1870. On withdrawal from service it was preserved at Swindon locomotive works until unforgivably broken up in 1906. Some atonement was made when a replica was built for the Stockton & Darlington Railway Centenary celebrations in 1925. This replica is now housed in the Great Western Railway Museum at Swindon in the Churchward gallery.

A replica of the *North Star* in the Great Western Railway Museum at Swindon (*University of Surrey*)

Between 1840 and 1842, over one hundred engines were built by various manufacturers to Gooch's designs including the Fire Fly, Sun, Leo and Hercules classes. No further engines were built for the Great Western Railway between 1842 and the opening of the locomotive works at Swindon a few years later. It was during this period that a Royal Commission was set up for the purpose of establishing a uniform 'Gauge'. Although the narrow gauge advocates had the undeniable advantage in terms of total mileage constructed, the combined talents of Brunel and Gooch presented a powerful case for the broad gauge. In 1845, Brunel proposed a series of tests (Macdermot) to compare the performance of broad and narrow gauge engines and *Ixion* of the Fire Fly class was chosen to represent the broad gauge interests.

The force that a locomotive can exert on the rails to provide traction is related to the boiler pressure (P lb/in^2), the cylinder diameter (d in), the piston stroke (S in) and the wheel diameter (D in). The tractive effort (T lb) is represented by the equation (see Technical Appendix):

$$T = \frac{0.425\ Pd^2S}{D}$$

Details of *Ixion* are as follows (P. J. T. Reed):

Wheel arrangement	2–2–2
Driving wheel diameter	7ft
Cylinder diameter	15$\frac{3}{4}$in
Piston stroke	18in
Boiler pressure	75lb/in^2

Substituting these figures in the above equation we have:

$$\text{Tractive effort } T = \frac{0.425 \times 75 \times 15.75^2 \times 18}{7 \times 12}$$

$$= 1{,}694 \text{ lb per cylinder}$$

In Table 7, the tractive effort of a number of Great Western engines and some early Stephenson engines is compared, indicating the former's superiority and the dramatic increase in power as design improved. The test course selected was the 53 miles between Paddington and Didcot. With loads between 60 and 80 tons, the 22-ton engine achieved a maximum speed in the trials in excess of 60mph. The general performance of the broad gauge engine was superior to the narrow gauge engine which ran on a trial course between York and Darlington. Although the Gauge Commissioners emphasised in their report the advantage of the broad gauge in terms of speed, it was recommended that the gauge of 4ft 8½in should be adopted throughout Great Britain.

In 1846, Gooch-designed engines emerged from the locomotive works at Swindon. Between 1846 and Brunel's death in 1859, about one hundred and fifty engines were built at Swindon and the first of these was the 2–2–2 *Great Western*. Details of this engine are given in Table 7 and it was the prototype for the famous Iron Duke class with 8ft driving wheels. The *Great Western* was subjected to extensive trials, and in June 1846, completed the 388-mile return journey from Paddington to Exeter at an average speed in excess of 55mph. On a shorter run to Swindon, it averaged just under 60mph pulling a load of 100 tons. The engine weighed about 30 tons and excessive weight distribution on the leading axle necessitated its reconstruction with a bogie. This modified 4–2–2 design was adopted for the twenty-nine engines of the Iron Duke class, twenty-two of which were built at Swindon. They were delivered between 1847 and 1851 and were renowned for their power and reliability. Fig 57 illustrates the general form of the Iron Duke class locomotive.

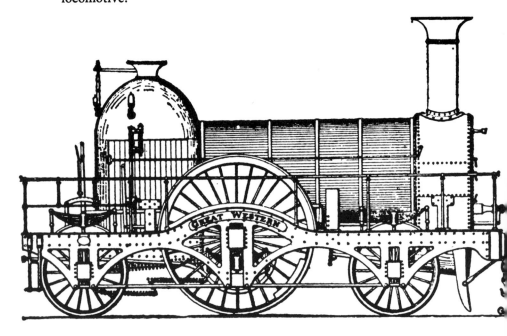

2-2-2 *Great Western*, the first GWR engine built entirely at Swindon, April 1846 (*University of Surrey*)

NAME & DATE	CLASS	WHEEL ARRANGEMENT	DRIVING WHEEL DIAM./FT	BOILER PRESSURE LB/IN^2	PISTON STROKE IN	CYLINDER DIAM. IN	TRACTIVE EFFORT LB	NO. TUBES IN BOILER	DESIGNER
ROCKET (1829)	—	0-2-2	4ft 8in	50	16½	8	801	25	G. and R. Stephenson
PLANET (1830)	—	2-2-0	5	50	16	11	1,370	129	G. and R. Stephenson
AJAX* (1838)	—	2-2-2	10	50	20	14	1,388	96	Mather, Dixon
NORTH STAR (1837)	Star	2-2-2	7	50-60	16	16	2,486	167	R. Stephenson
IXION (1841)	Fire Fly	2-2-2	7	50	18	15	3,388	131	D. Gooch
GREAT WESTERN (1846)	—	2-2-2 (originally)	8	100	24	18	6,884	278	D. Gooch
LORD OF THE ISLES (1851)	Iron Duke	4-2-2	8	120	24	18	8,262	303	D. Gooch

Table 7 Technical details of five GWR engines built between 1837 and 1851. Details of the Rocket and Planet are included for comparison *Manufactured by Mather, Dixon to Brunel's specification.

Fig 57 4–2–2 Iron Duke class locomotive with 8ft driving wheels

The last of these engines to be built was *Lord of the Isles* and it worked between 1852 and 1884. Its total mileage exceeded three quarters of a million. Between 1840 and 1850 Gooch continuously improved the design of his engines and carried out a number of experiments to estimate train resistances. It was during this period that Brunel's ambition of constructing the best railway that imagination could devise approached reality. The powerful Gooch-designed engines built and serviced in Brunel's splendid structures at Swindon could take full advantage of the shallow gradients between Paddington and Exeter. Brunel freely acknowledged the contribution of the Stephensons and Daniel Gooch to the success of the locomotive division. A fleeting impression of the early days of the Great Western Railway is given in J. M. W. Turner's (1775–1851) painting 'Rain, Steam and Speed, The Great Western Railway' (1844). It depicts a Gooch-designed 2–2–2 Fire Fly class crossing Maidenhead Bridge and travelling westward.

It was on the Great Western that Queen Victoria made her first railway journey. On Monday 13 June 1842, the royal train, consisting of the royal saloon and six other carriages and trucks, was worked by the 2–2–2 *Phlegethon* of the Fire Fly class with Gooch driving and Brunel at his side. The 18-mile journey was completed in 25 minutes. The Queen was pleased with the experience and subsequently made frequent journeys to Slough where two reception rooms were appropriately furnished for her use. The spur to Windsor was not opened until October 1849.

It appears that Brunel did not drive locomotives (Macdermot) and in 1841 told the Parliamentary Committee on Railways: 'I never dare drive an engine, although I always go upon the engine; because if I go upon a bit of line without anything to attract my attention I begin thinking of something else.'

10
TUNNELS

Tunnel is now, I think, dead. The Commissioners have refused on the ground of want of security. This is the first time I have felt able to cry at least for these ten years. Some further attempts may be made – but – it will never be finished now in my father's lifetime I fear. However, nil desperandum has always been my motto – we may succeed yet. Perseverantia.   (I. K. B. 1831)

THE THAMES TUNNEL

The Thames Tunnel was not opened as a public thoroughfare until March 1843, over eleven years after Brunel's somewhat depressing entry into his diary in December 1831. It was built between 1825 and 1843 and may be considered as one of the most hazardous civil engineering projects undertaken during the first half of the nineteenth century.

Professor W. J. M. Rankine (1820–72), in his *Manual of Civil Engineering*, described the Thames Tunnel works under the heading 'Tunnelling in mud' (I. Brunel). Rankine, who held the chair of engineering at Glasgow University between 1855 and 1872, made an immense contribution to the development of a scientific approach to the understanding of the stability of soils and was obviously well aware of the difficulty of excavating a tunnel 37ft 6in wide, 22ft 3in high and 1,200ft long through soft soil a short distance below the bed of the river. In fact, the difficulties encountered almost defeated the combined talents of the Brunels and it is not surprising that two previous attempts to tunnel under the Thames, made in 1801 and 1807, were unsuccessful. The second attempt was under the direction of Richard Trevithick (see also Chapter 9), and a pilot tunnel or heading was driven over 1,000ft from a shaft sunk at Rotherhithe on the south bank of the Thames opposite Wapping.

However, the excavation met quicksand, a sand through which water moves so fast that the sand is held in suspension by the water, and consequently the tunnel was flooded. Work on the project was abandoned, and it was then the general opinion in engineering circles that there was not a practical means of tunnelling under the Thames. However, a patent specification filed by Marc Isambard Brunel in January 1818, 'Forming Drifts and Tunnels Under Ground' led, some twenty-five years later, to the opening of the first *sub acqueous* tunnel built for public use.

A few years previously, Marc Isambard Brunel was working on the sawmill project at Chatham Dockyard (see Chapter 1) and it is said that his ideas for developing a tunnelling shield, which supported the excavation until the lining of brickwork was completed, were developed from observations of the *Teredo navalis* (ship worm) whose encased head and soft body could penetrate the hardest oak of a ship's hull (Clements).

The tunnelling shield, which was described by Marc Isambard as 'an ambulating cofferdam, travelling horizontally', was required to support a rectangular excavation, which was brick lined and pierced by two horseshoe arches, also pierced, to link the two passageways. The operation of the shield will be described prior to considering the overall progress of the construction of the tunnel. A section through one of the frames of the shield is shown in Fig 58. There were twelve vertical cast-iron frames (1) side by side, each 3ft wide and 22ft high which occupied the full width of the excavation. Each frame contained three cells within which the miners would work. The men were protected from the earth in front by a series of horizontal boards (poling boards) (2) which were held

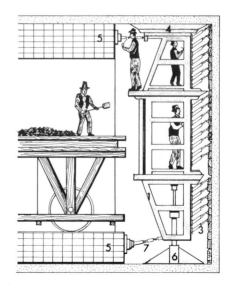

1 cast iron frame
2 poling boards
3 screw jacks holding poling
   boards in place
4 iron plates
5 brick tunnel lining
6 pivoted feet to frames
7 main jacks

A section through Marc Isambard Brunel's tunnelling shield, patented in 1818.

Fig 58

in place by a series of screw jacks (3). The earth at the sides and top of the frames was held in place by a series of iron plates (4). The frames themselves were jacked against the completed brick lining (5).

Each poling board was removed separately and the earth excavated a few inches, the board replaced and jacked against the newly formed face. When the entire face had been excavated, the cast-iron frames were moved forward in the following manner. Each frame pivoted on two feet (6) and each foot was lifted up in turn and moved forward and screwed down onto the ground. By releasing the screw jacks supporting the poling boards it was possible to move each frame forward by means of the main jacks (7) located at the top and bottom of the frames. The iron plates at the sides and top of the frames were moved forward separately.

In the spring of 1823 Marc Isambard and Isambard Kingdom were working on a scheme for constructing a double roadway under the Thames between Rotherhithe and Wapping about two miles below London Bridge, and with the benefit of the tunnelling shield were able to attract financial support. In the following year, the Thames Tunnel Company was incorporated by an Act of Parliament with Marc Isambard as engineer.

The Brunels were aware of the importance of establishing the geological characteristics of soils, and Marc Isambard employed engineers to make thirty-nine borings in two parallel lines across the river in an attempt to establish the characteristics of the soil beneath the Thames. Engineering geology was in its infancy as a scientific study and the results obtained were by no means reliable, but he was certainly aware of the presence of the quicksand encountered by Trevithick. The intention was to drive the tunnel through a stratum of relatively impermeable firm clay.

In order to commence the tunnelling at the preferred level and to accommodate the shield, it was necessary to construct a 50ft diameter vertical shaft on the Rotherhithe bank. Work commenced in February 1825, and the 3ft-thick brickwork lining reinforced with vertical metal rods was built up to a height of about 40ft. The shaft was rendered externally with a sand/cement mix and sunk under its own weight as the men excavated below the bottom edge. In order to provide an opening for the tunnelling shield the remaining depth of 25ft was constructed by adding sections of brickwork below in stages known as underlaying or underpinning. This was a tedious process and for the Wapping shore Marc Isambard resolved to sink the shaft to its full depth without any underpinning. A reservoir with a domed roof was constructed at the bottom of the shaft to accommodate permanent pumping machinery. A section through the shaft and geological data are shown in Fig 59. Construction of the shaft extended over a period of eight months and work was frequently interrupted by the ingress of water.

Tunnelling operations commenced in November 1825 and it was estimated that the works would be completed in three years. Due to Marc Isambard's illness the works were left under the direction of his son. Isambard Kingdom played a vital role in the day to day supervision of the tunnelling work and it is difficult to conceive a more arduous project on which to gain practical experience. An indication of the slow progress of the works is given in Table 8. (I. Brunel)

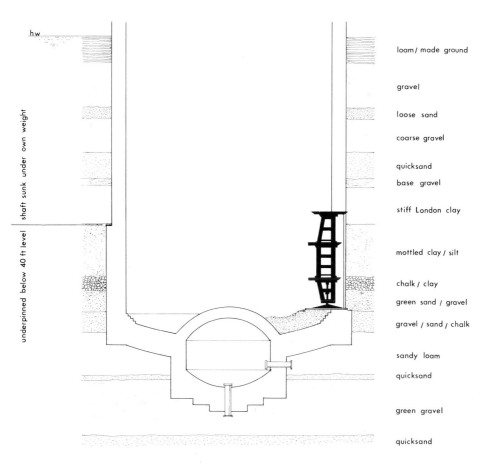

Fig 59 A section through the Rotherhithe shaft showing the tunnelling shield in position and geological data. Shaft underpinned below 40ft level

DATE	LENGTH COMPLETED FEET
Nov. 1825	0
May 1826	100
Aug. 1826	205
Dec. 1826	350
Feb. 1827	405
Jan. 1828	600

Table 8 Progress of the tunnelling shield between November 1825 and January 1828

Until April 1826, the resident engineer was William Armstrong but he resigned due to ill health and his place was taken on a acting basis by Isambard Kingdom. Isambard Kingdom devoted all his energy to maintaining the progress of the work, and numerous entries in Marc Isambard's diary indicate his appreciation of his son's efforts and concern for his health:

Isambard's vigilance and constant attendance were of great benefit. He is in every respect a most useful coadjutor in this undertaking. (5 June 1826)

Isambard was the greater part of the night in the works and the benefit of his exertions is indeed most highly felt: no one has stood out like him. (18 September 1826)

A work that requires such close attention, so much ingenuity and carried on day and night by the rudest hands possible – what anxiety, what fatigue, both of mind and body. Every morning I say, Another day of danger over. [On 3 January 1827 Isambard Kingdom was formally appointed resident engineer.] (4 January 1827)

In November 1826 Isambard Kingdom entered in his diary that he had spent for nine days on an 'average $20\frac{1}{3}$ hours per day in the tunnel and $3\frac{2}{3}$ hours to sleep'.

On 18 May 1827, the river broke in for the first time and Isambard Kingdom made a number of descents in a diving bell to inspect the breach in the river bed. It was sealed by filling the hole with bags of clay and gravel. After the fill material had consolidated, excavation was continued, but on 12 January 1828, a second major flooding occurred and Isambard Kingdom was seriously injured during the rescue operations in which he saved the lives of several men. The tunnel was cleared of water and the directors ordered the end of the tunnel adjacent to the shield to be bricked up. It was not possible to raise the money to continue the work until 1835 when, with the aid of a loan from the Government, a new shield with a number of improvements was substituted.

The river penetrated the tunnel on three subsequent occasions, and in October 1840 work was commenced on the Wapping shaft which was sunk, without underpinning, to a depth of 70ft. The shield was then brought up to the shaft and the connection between the tunnel and the shaft made. It was opened to the public on 25 March 1843 and over the years attracted millions of visitors. A cross-section through the tunnel is shown in Fig 60. The proposed approaches for vehicular traffic were

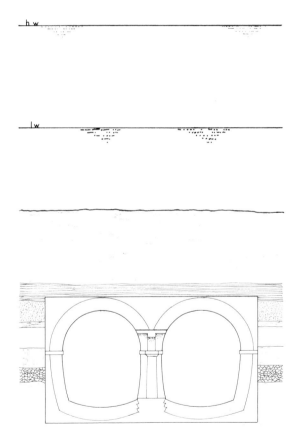

Fig 60 A cross section through the Thames Tunnel showing the twin horseshoe passageways within the rectangular excavation

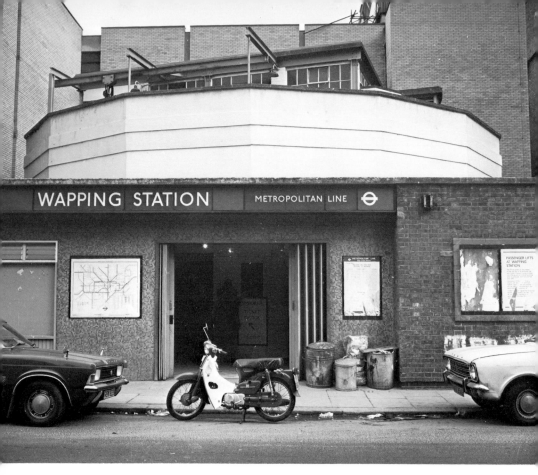

The entrance to Wapping Station built at the site of the original shaft (*Frances Gibson Smith*)

never built. The receipts were minimal when compared with the total cost of over £460,000, and in 1865 the tunnel was sold to the East London Railway for £200,000 (Lee). The East London line now forms part of the London Underground network and connects Shoreditch with New Cross, with intermediate stations at Whitechapel, Shadwell, Wapping, Rotherhithe and Surrey Docks. Wapping Station was built at the site of the original shaft and the lifts are built on an iron framework within the shaft.

From the south end of the platform it is possible to observe the twin horseshoe tunnels dipping under the Thames. In the entrance to Wapping Station a plaque was erected to commemorate the centenary of the death of Isambard Kingdom Brunel which reads:

The tunnel which runs under the Thames from this station was the first tunnel for public traffic ever to be driven beneath a river. It was designed by Sir Marc Isambard Brunel 1769–1849 and completed in 1843. His son Isambard Kingdom Brunel 1806–1859 was engineer in charge from 1825 to 1828.

Rotherhithe Station was built a short distance to the south of the tunnel shaft on the Surrey bank, and the area to the north of the station is designated St Mary's Rotherhithe Conservation Area and embraces the tunnel shaft, Brunel's engine house, the Mayflower Inn, the Thames Tunnel Mills, Hope Sufferance Wharf and St Mary's Church. The engines required to work the pumps and remove excavated material from the tunnel were initially located over the shaft, but on completion of the works, a separate engine house was built a short distance to the west of the shaft. The engine house of plan dimensions 20ft by 40ft, with a tapered square chimney, was constructed about 1842 and the roof has recently (1979) been restored in order to provide a space to exhibit a permanent display on the Brunels, the Thames Tunnel and the history of Rotherhithe.

Marc Isambard Brunel was rewarded with a knighthood for his services to the construction of the Thames Tunnel in March 1841, and his son's reward was the unrivalled practical experience gained from the age of nineteen to twenty-two as 'engineer-in-charge' of the works. Isambard Kingdom's association with the tunnel ceased in 1828 and shortly there were other matters to receive his attention – Clifton Suspension Bridge and the Great Western Railway. The route chosen for the Great Western Railway involved the construction of eight tunnels between Chippenham and Bristol, the first of which, about 6 miles west of Chippenham, was the 9,600ft long tunnel at Box.

BOX TUNNEL

In conclusion, I must observe that no man can be more sensible than I am of the great advantage it would be to me as a civil engineer to be better acquainted with geology, as well as with many other branches of science, that I have endeavoured to inform myself on the subject, and that I have not altogether thrown away the many opportunities afforded me in my professional pursuits.

(I. K. B. June 1842)

The above is an extract from a letter to an eminent geologist who had expressed doubts as to the safety of Box Tunnel, written about a year after its opening in June 1841. It expresses clearly Brunel's view on the need for catholic knowledge in engineering matters and the necessity to use previous experience. The letter also dwells at some length on his detailed examination of samples of rock through which the tunnel was driven:

I cannot have gone through such a study without acquiring a very intimate and practical knowledge of the structure and the peculiarities of the particular mass of rock which is now in question; and I will say frankly what I feel upon this point, which is, that I ought now to possess a more thorough and practical knowledge of this particular rock and its defects, and the best mode of remedying them, than even you yourself, with your immeasurably greater scientific knowledge of rocks generally.

Whilst putting the eminent geologist gently in his place, Brunel's letter emphasises the importance he attaches to the application of geology to civil engineering problems.

Brunel's previous experience under his father's guidance on the construction of the Thames Tunnel was of course invaluable, but the ground conditions at Box were completely different. It was proposed to excavate the tunnel at a gradient of 1 in 100 falling westwards through Box Hill, and as we have seen (Chapter 3) this proposal met with considerable opposition, although a number of canal tunnels of similar or greater length including Norwood 1.8 miles and Sapperton 2.2 miles had been constructed before the end of the eighteenth century.

Geological map No 265 'Bath' published by the Ordnance Survey, Chessington and other sources (Pugsley *et al.*) reveal that the tunnel passes through a number of strata (Fig 61) – forest marble, great oolite, fuller's earth, inferior oolite and lias clay. The forest marble is a rock which prior to being recrystallised was limestone, that is, mainly calcium carbonate. The great and inferior oolites are also limestones and the term oolite refers to a rock consisting of small rounded particles. There are many varieties of limestone, the best known of which is chalk. Fuller's

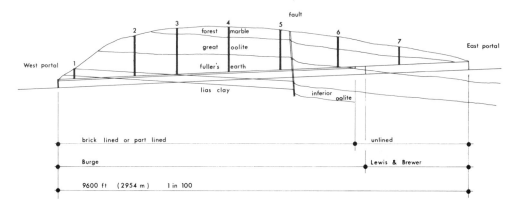

Fig 61 A longitudinal section through Box Tunnel showing the ground strata

earth is a clay material which has important industrial applications due to its absorbent properties. The term lias relates to a period in the earth's history known as the Jurassic period which occurred in the Mesozoic (middle life) era.

The first stage of the work, which consisted of sinking a series of temporary shafts along the line of the tunnel to determine the position of the various strata, was commenced at the beginning of 1836. This was followed by the main shafts numbered 1 to 8 from west to east. The outer shafts 1 and 8 were subsequently opened up into cuttings at each entrance to the tunnel. Shafts 2 to 7 are shown in Fig 61 and their depths were 285ft, 293ft, 283ft, 250ft, 230ft and 125ft respectively. They were generally 25ft in diameter and lined with brick or masonry about 2ft thick. The shafts served the dual purpose of providing ventilation and as a means of removing material from the tunnel. The material was removed from three of the shafts by means of steam power (Whishaw) and for the remainder a horse gear or gin was employed. This consisted of a lever drawn round a circle by two or three horses, the lever providing the means of rotating a double drum round which ropes were wound and connected to buckets. As the drum rotated, one bucket descended as the other ascended. The horses walked in a circle of not less than 24ft diameter.

Excavation of the tunnel itself commenced in the spring of 1838 and in all about 247,000 cubic yards were removed. Contracts were let to Lewis & Brewer who were responsible for the first 2,700ft from the east portal, and to George Burge, who was responsible for the remaining 6,900ft. Lewis & Brewer's work required excavation through the great oolite and it is stated that up to a ton of gunpowder was used each week. It was considered that the great oolite would be self-supporting and this section of the tunnel was not lined. A lithograph by J. C. Bourne shows an unlined section of the tunnel. The overall dimensions of the excavation were approximately 30ft wide and 36ft high.

It was necessary for Burge to excavate in material which required support, although it was not considered necessary to utilise a tunnelling shield of the type patented by Marc Isambard Brunel and a less sophisticated procedure was adopted which had been developed in the canal building period. The excavation was lined with six or seven rings of brickwork. At the end of February 1840, Brunel reported that 5,700ft of the tunnel had been excavated and over the previous three months the rate of progress was about 6ft per day. The best rate of progress during the construction of the Thames Tunnel was 14ft per week and it was generally less than half that. Construction of the tunnel involved round

J. C. Bourne's lithograph showing an unlined section of Box Tunnel

the clock working and progress was hindered by the ingress of water. Additional shafts were sunk and during the final six months it is said that 4,000 men and 300 horses were at work on the project (Macdermot). The tunnel was completed in June 1841.

The remaining tunnels between Box and Bristol were much shorter in length, the largest of which was the 3,150ft Brislington Tunnel about half-way between Keynsham and Bristol Temple Meads. The excavation of this tunnel required the construction of the three main shafts for removal of material and six smaller shafts for ventilation purposes. Other tunnels constructed under Brunel's general direction were Whiteball Tunnel (3,276ft) on the Bristol & Exeter Railway, and Sapperton Tunnel (5,580ft) on the Cheltenham & Great Western Union Railway.

Over a period of twenty years, Isambard Kingdom Brunel was involved in the construction of tunnels in strata which ranged from mud to rock, yet even at the height of his involvement in the day by day hazards of the construction of the Thames Tunnel, he found time for a little dalliance: 'I have made love openly to C. H. and received a return, and doing the same thing with M. C. here – she has made a fool of me, and I of her.' (Clements)

11
STEAMSHIPS

And in the throbbing engine room
Leap the long rods of polished steel . . . (Oscar Wilde)

In the period of just over one hundred years between the launching of England's most famous sailing ships, HMS *Victory* in 1765 and the *Cutty Sark* in 1869, steam power and iron hulls became well established as an alternative to sails and wooden hulls. A major contribution to this achievement was made by Isambard Kingdom Brunel in the design and construction of three steamships the PS *Great Western* (1837), the SS *Great Britain* (1843) and the PSS *Great Eastern* (1858). All three, in their time, were the world's largest ocean-going passenger ships.

The construction of HMS *Victory* at Chatham, requiring over two thousand oak trees to be built, was well before Marc Isambard's involvement in the sawmill, but she is now preserved for posterity in Portsmouth Dockyard less than a mile from his son's birthplace. It was in Portsmouth that Marc Isambard became involved in the design of block-making machinery, and no doubt some of the pulley blocks which came off the production line were used in refits of the *Victory*. The tea clipper, *Cutty Sark* is in dry dock at Greenwich, close to the National Maritime Museum and a short distance downstream from the site of the launching of the *Great Eastern*. Fortunately, one of Brunel's steamships, the *Great Britain*, still exists and lies in the Great Western dry dock at Bristol, undergoing restoration.

The structural design of a ship has much in common with that of bridges and the force actions described in Chapter 6 should be considered when proportioning a ship structure. However, the terminology used in ship design is different from that generally adopted for bridges (see Fig 62). Hogging and sagging bending effects are related to the position of the ship's hull in the wave profile; it can twist or rack in heavy seas and provision should be made for the stresses induced on grounding (see section on SS *Great Britain*). A concentration of heavy loads along the

P.S. "Great Western"
1837

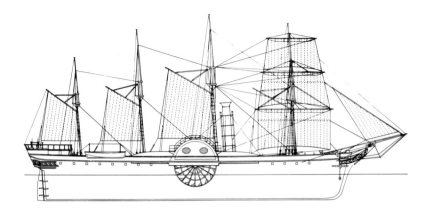

S.S. "Great Britain"
1843

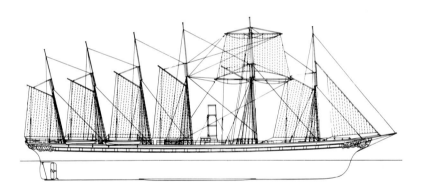

P.S.S. "Great Eastern"
1858

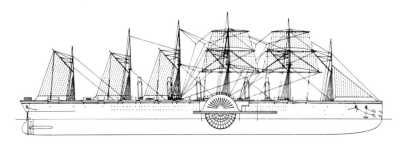

Brunel's three steamships

centre of the hull induces deformations which tend to draw the upper
sides of the ship together, known as collapsing stresses. The first person
to treat the hull of a ship as a beam was possibly Thomas Young (see
Chapter 6). He assumed a distribution of the weights of the different
portions of the hull, defined wave profiles and calculated the bending
effects at various sections. In a report to the Board of Admiralty, he
indicated that it would be difficult to unite a series of parallel planks by
pieces crossing them at right angles so that they would not slide relative
to each other.

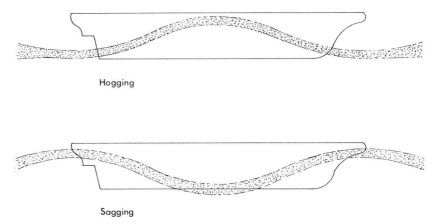

Hogging

Sagging

Fig 62 Hogging and sagging bending effects in relation to the wave profile

To prevent this sliding motion, Young suggested the adoption of
diagonal bracing in wooden ships. The cross and longitudinal sections
through HMS *Victory* (Fig 63) show that she was constructed in the
conventional manner of the latter half of the eighteenth century, without
diagonal bracing. The keel was constructed from several lengths of wood
which were overlapped and bolted together, and the stem and sternpost
fixed at each end. The frames forming the bottom and the sides of the
ship were then placed at right angles to the keel and clamped to it by the
keelson. The resulting framework was then given two skins of planking
at the sides and this was followed by planking the decks. The ship was
coppered below the waterline as a protection against the *Teredo* worm. It
was not until the second decade of the nineteenth century that Sir
Robert Steppings, Surveyor to the Navy from 1813–32 introduced
diagonal bracing into the sides and decks of ships' hulls (Fig 64). At the
time of the construction of the *Cutty Sark*, the iron hull was well
established (see SS *Great Britain* and PSS *Great Eastern*), but in

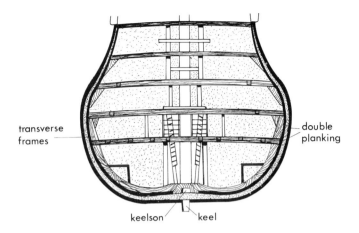

Fig 63 Longitudinal and cross-sections through HMS *Victory*

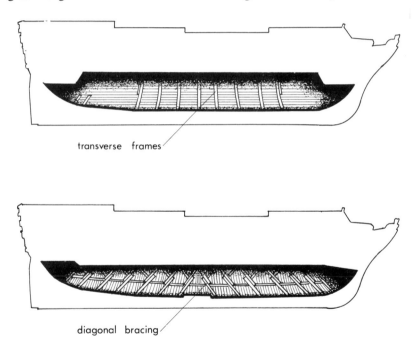

Fig 64 Diagonal bracing introduced to strengthen the hull from about 1810 onwards

tropical waters fouling of the hull reduced efficiency. Due to galvanic corrosion, it was not possible to sheath iron plates with copper, and a composite construction was adopted for clipper ships. A typical cross-section is illustrated in Fig 65. The interior frames, deck beams, columns, diagonal bracing and keelson were of iron and the keel, stem,

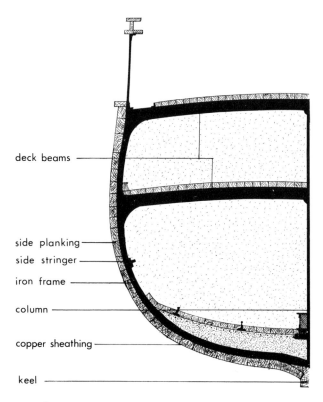

deck beams

side planking

side stringer

iron frame

column

copper sheathing

keel

Fig 65 Features of a composite construction of clipper ships. Diagonal ties were also fixed to the side planking

stern and planking were of wood sheathed by copper plating below the deck line. The hull structure of Brunel's steamships will be described later.

In addition to consideration of the strength of the hull structure, two further essential aspects of the design of a ship are its stability when floating in water and the hull form in relation to the way in which it cuts through the water. The principle of ship stability is given a brief quantitative treatment in the Technical Appendix.

Turning to the form of a ship, it is necessary to define some basic geometrical concepts by means of Fig 66. The term 'block coefficient' is used as a measure of the fullness of form of a ship and is defined as the ratio of the volume of displacement ∇ to a given water line to the volume of the circumscribing solid of constant rectangular cross-section having the same length L_{pp}, breadth B and draught T as the ship (see Fig 66).

$$\text{Thus block coefficient } C_B = \frac{\nabla}{L_{pp} \times B \times T}$$

Typical values of the block coefficient range from 0.5 for fast ships (fine form) to 0.88 for slow cargo ships (full form). The volume of displacement is the total volume of water displaced by the ship. Table 9 compares the principle dimensions and other details of HMS *Victory*, the *Cutty Sark* and Brunel's three steamships. The values listed for the block coefficients are only approximate and based on salt water density of 64 pounds per cubic foot. They show that all five ships can be classified as having a relatively fine form.

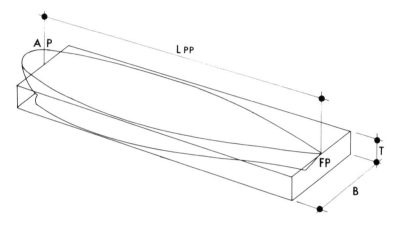

Fig 66 L$_{pp}$ is length between the forward perpendicular FP and the after perpendicular AP

EARLY STEAMBOATS

In the last quarter of the eighteenth century, a number of experiments were made with marine steam engines and propulsion by means of water jet, paddles and screw propeller. In 1783, the Marquis de Jouffray operated a 180-ton wheel steamer on the River Saône near Lyons in France, and a few years later, John Fitch, an American inventor, was experimenting with various types of steamboat, including one with a screw propeller. At the turn of the century, William Symington (1763–1831) patented an improved steam engine in which a connecting rod from a horizontal cylinder drove direct onto a cranked paddle shaft. He fitted up a steam tug, the *Charlotte Dundas* and this was possibly the first successful power driven vessel. An American witness to Symington's work was Robert Fulton (1765–1815), who built a passenger boat 150ft long with Boulton and Watt engines (see Chapter 9) powering 15ft diameter paddles. In 1807, this vessel completed the 150-mile run from New York to Albany along the Hudson River in 32 hours. In 1811,

DETAIL	HMS VICTORY	PS GREAT WESTERN	SS GREAT BRITAIN	PSS GREAT EASTERN	CUTTY SARK
Launch date	1765	1837	1843	1858	1869
Length overall ft(m)	226.5 (69.7)	236 (72.6)	322 (99.1)	692 (212.9)	280 (86.2)
Length between perpendiculars ft	190	212	285	680	212.5
Displacement at load draught tons	3,500	2,300	3,675	27,380	2,100
Breadth of hull ft (m)	51.8 (15.9)	35.3 (10.9)	50.5 (15.5)	82.5 (25.4)	36 (11.1)
Approximate draught laden ft (m)	22 (6.8)	16.7 (5.1)	18 (5.5)	30 (9.2)	20 (6.2)
Hull material	Wood	Wood	Iron	Iron	Iron/Wood
Block coefficient	0.57	0.64	0.5	0.57	0.48
Built	Chatham	Bristol	Bristol	Millwall	Dumbarton

Table 9 Principal dimensions and other details of HMS *Victory*, *Cutty Sark* and Brunel's three steamships (metric units in brackets)

Henry Bell, who was influenced by the work of Symington and Fulton, launched a small passenger steamer, the *Comet*, powered by a 3hp engine which was the forerunner of fleets of small steamships which plied the Clyde, Firth of Forth and the Thames.

In the first decade of the nineteenth century, iron-hulled ships began to replace wooden vessels, and in the 1820s William Fairbairn (1789–1874), who was later associated with Robert Stephenson in the design of the Britannia Tubular Bridge (see Introduction), built four iron ships, prior to moving south and establishing a shipbuilding yard at Millwall, London.

The first crossing of the Atlantic, partially assisted by steam power, was in 1819 by the 320-ton *Savannah* built in New York as a fully rigged sailing ship and then fitted with auxiliary engines. Eight years later, the 438-ton Dutch owned and British built wooden paddle ship, the *Curaçao* made the first all-steam Atlantic crossing. In the 1830s a number of steam powered ships crossed the Atlantic, but the initiation of a regular transatlantic passenger service was made by the 2,300-ton wooden-hulled PS *Great Western* designed by Isambard Kingdom Brunel.

PS *GREAT WESTERN*

Why not make it longer and have a steamboat to go from Bristol to New York, and call it the Great Western   (I. K. B. 1835)

In the above quotation, 'it' refers to the Great Western Railway, and Brunel's proposal to initiate a combined rail and sea transportation system from London to New York was soon taken up and thus began his involvement with ship design, on a scale that dwarfed any previous achievements. It has been demonstrated that innovations such as the marine steam engine, treatment of a ship's hull as a beam, the use of diagonal bracing, iron hulls and screw propulsion had all been developed prior to 1835. In terms of scale, Brunel's contribution to naval architecture is irrefutable and the PSS *Great Eastern* remained the world's largest ship until the end of the nineteenth century. In designing ships of such large dimensions (see Table 9), Brunel was faced with two major problems – the necessity for increased longitudinal strength to resist the action of heavy Atlantic waves and the evaluation of the fuel capacity required to complete a passage of several thousand miles. With the formation of the Great Western Steamship Company in 1836 with Brunel as a member of the building committee, the opportunity for attempting to solve these problems presented itself.

The structural design of the wooden-hulled PS *Great Western* was fairly conventional – the oak frames or ribs forming the bottom and sides of the ship were closely spaced and dowelled and bolted together in pairs, and she was trussed with iron and wooden diagonals to resist racking forces. An innovation was the introduction of four staggered rows of $1\frac{1}{2}$in diameter iron bolts which ran longitudinally through the length of the ship to increase its longitudinal strength. The bolts were 24ft long and lapped to provide continuity. The hull was sheathed in copper below the water line.

The question of fuel capacity and engine size was more controversial. From a fact-finding tour of British ports, two members of the committee, T. P. Guppy and Captain C. Claxton (managing director of the company), were able to prepare a report which included a passage from Brunel claiming that the capacity of a hull increased with the cube of its dimensions but the power required to drive it through water only increased with the square of these dimensions. However, opinion persisted that the coal required to propel a steamship would increase as a cube of its dimensions and Dr Lardner, who opposed a number of Brunel's projects, produced calculations at a meeting of the British Association for the Advancement of Science, which demonstrated that a transatlantic steamer would require a relay of coals. Brunel detected errors in Dr Lardner's calculations and considerable doubt still remained with regard to a crossing without substantial assistance from sails.

The dilemma was soon to be put to the test as the PS *Great Western* was launched on 19 July 1837 and her 420hp engines were built and fitted in London by Messrs Maudslay, Sons & Field. The return journey to Bristol was commenced on 31 March 1838 with Brunel, Guppy and Claxton on board. About two hours after the commencement of the journey a fire broke out and Brunel suffered serious injury. This delayed the ship's arrival at Bristol and it was not until 8 April 1838 that the *Great Western* set out for New York with a handful of passengers. Three days previously, a rival ship, the London to Cork paddle steamer, the 700-ton *Sirius*, left Cork for New York and arrived in the early morning of 23 April 1838 having consumed all her coal and some of the fittings. Twelve hours later, the *Great Western* arrived having bettered *Sirius*'s time by nearly four days, and with coal to spare.

The *Great Western* became the 'Queen' of the Atlantic running regularly between Bristol and New York between 1838 and 1846. Prior to the construction of the *Great Western*, the United States dominated the North Atlantic passenger trade with fast sailing ships (mail packets) of up to 500-tons displacement and this was the means by which Marc

Brunel had crossed the Atlantic some forty years previously. With favourable weather conditions, the sailing ships' shortest crossing time was as low as fourteen days but could double in adverse conditions. In 1847, the *Great Western* was purchased by the West India Mail Steam Packet Company and remained in service for a further twenty years until, in 1857, she was broken up in London.

SS *GREAT BRITAIN*

On 19 July 1970 the hull of the SS *Great Britain*, ravaged by the sea for one hundred and twenty-seven years, but still afloat on her own bottom, entered the Great Western dry dock where she had been built between 1839 and 1843. The many phases of the life of this ship, her salvage and the progress of restoration are described in Dr Ewan Corlett's fascinating and superbly illustrated book *The Iron Ship*. The restoration work, due to be completed about 1986, is being carried out under the auspices of the SS *Great Britain* Project Company. Following a brief

The SS *Great Britain* in the Great Western dry dock at Bristol undergoing restoration (*Frances Gibson Smith*)

description of the overall life of the ship, the first ten years, her original Brunel period, will be considered in greater detail.

The *Great Britain* took four years to build and was first floated in the Great Western dry dock on 19 July 1843. It was not until December of the following year that the 1,000hp engines driving a $15\frac{1}{2}$ft diameter six-bladed propeller were put to the test. A speed of 11 knots was achieved and official trials commenced in January 1845.

Her maiden voyage to New York from Liverpool commenced in July 1845, and on her second voyage the propeller sustained extensive damage. A winter refit included replacing the six-bladed propeller by a four-bladed one, modifications to the boilers and reducing the number of masts to five. In the early summer of 1846, voyages to New York were recommenced and on her fifth voyage she ran aground on the beach of Dundrum Bay in County Down on the east coast of Ireland. Details of the salvaging operation will be described later and it was not until the end of August 1847 that she was towed back to Liverpool for repairs. The Great Western Steamship Company was wound up, and the *Great Britain* was sold and underwent an extensive refit. The engines were replaced, the number of masts was reduced to four, twin funnels were installed and an upper deck house added to increase accommodation. She was intended for service between England and Australia and her first voyage took 83 days with a crew of 142 and 630 passengers. There were further refits between 1852 and 1857, the number of masts was reduced to three with square rigging, the single funnel returned and a two-bladed propeller installed. Although mainly used for the Australia run, the *Great Britain* was also used as a troopship with trips to the Crimea and Bombay. In 1876 she came out of passenger service and was eventually converted for use as a cargo sailing ship, the engines being removed and the hull strengthened by adding pine cladding between the low and high loading marks.

Ten years later, with her ageing hull laden with coal, the *Great Britain* was unable to round Cape Horn in heavy seas and found shelter in Port Stanley in the Falkland Islands. She was bought as a store ship and remained in Port Stanley for fifty years. In 1937 she was towed to nearby Sparrow Cove, beached, holed and left to disintegrate, but this was not to be. In 1970 she was refloated, and towed by means of a pontoon 7,000 miles back to Avonmouth, then towed on her own bottom up the Avon passing under the Clifton Suspension Bridge through the locks and into the Floating Harbour. It is the intention that the restoration will follow closely Brunel's original design.

Returning to the summer of 1838, following the success of the *Great*

Western, the directors of the Great Western Steamship Company decided to build a second steamship. It was originally intended that the second ship would be a larger version of the *Great Western* and built from wood. However, the progress of iron ship design has been mentioned previously and no doubt this influenced Brunel who, in 1838, prepared calculations comparing the cost and efficiency of iron and wooden vessels. He also received a report from members of the building committee on the performance of an iron ship the *Rainbow*, based on observations during a voyage to Antwerp and back. The report favoured the use of iron and thus it was decided to alter the specification. A number of designs were considered and at the fifth attempt it was finally decided to go ahead, and on 19 July 1839 building commenced. The adoption of longitudinal stiffening bolts in the *Great Western* demonstrated Brunel's awareness of the beam action of long hulls and the following extract from a letter to Mr Guppy describes his ideas on the design of iron ships:

I have been thinking a great deal of your plans for iron ship building, and have come to a conclusion which I believe agrees with your ideas, but I will state mine without reference to yours. At the bottom and at the top I would give longitudinal strength and stiffness, gaining the latter from the former, so that all the metal used should add to the longitudinal tie, while in the neutral axis and along the sides, and to resist swells from the seas, I would have vertical strength by ribs and shelf pieces, thus: the black lines [Fig 67] being sections of longitudinal pieces, the dotted lines vertical and transverse diagonal plates, throwing the metal as much as possible into the outside bottom plates, and getting the strength inside by form, that is depth of beams etc, the former being liable to injury from blows, etc, the latter being protected. (I. K. B. 1843)

The idea of strength through form is clearly expressed and of course was applied on a much larger scale than attempted previously. The overall dimensions of the *Great Britain* are given in Table 9. The essential features of the hull construction are shown in the longitudinal and transverse section in Fig 67. The bottom was of cellular construction consisting of ten longitudinal members plated top and bottom. The frames or ribs were at close but uneven spacing to which the horizontal deck members were fixed. The deck members were supported internally by columns and at the junction with the side framing, diagonal ties were provided. At the top three decks, 3ft wide shelf plates were added together with Baltic pine longitudinal ties. Transversely the ship was partitioned into six compartments by five watertight bulkheads (originally used in Chinese junks), but these did not give complete protection as they were not carried up to the underside of the top deck. In

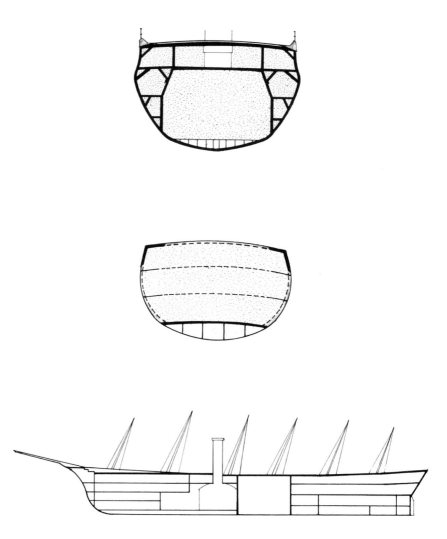

Fig 67 Hull construction of SS *Great Britain*

all, this was an efficient structure, exemplified by her long life, but complete cellular action could not be achieved without iron plating to the top deck. Her block coefficient, a measure of her fullness of form (see Table 9) is in the order of 0.5. Her external plates were lapped in the manner of wood clinker building. The plate size was 72in by 34in and they were double riveted to each other. The wrought-iron plates were manufactured by the Coalbrookdale Company at the Horsehay Ironworks and then transported via barges along the River Severn to Bristol.

Apart from the structural details of the ship, Brunel was also involved in the decision to adopt a screw propeller instead of a paddle wheel. In

October 1840, he submitted a lengthy report to the directors of the Great
Western Steamship Company which gave details of experiments carried
out on a 237-ton vessel, the *Archimedes*, with a 90hp engine driving a
$5\frac{3}{4}$ft diameter screw propeller which was patented by Francis Pettitt
Smith. It was concluded from these experiments that the screw was
superior to the paddle as a means of propulsion, but both its advantages
and disadvantages were considered. The report answers the objections
and then goes on to list six principal advantages including reduction in
weight, improved hull form and steering, regularity of motion and
adaptability of the power of the engine to changing weather conditions.
In the autumn of 1840 a resolution to adopt the screw propeller was
passed and the work on the original engines for the paddle wheel design
was abandoned. The original $15\frac{1}{2}$ft diameter propeller was six bladed
and although efficient in terms of propulsion, it was structurally weak
and replaced after two Atlantic voyages by a four-bladed design (see Fig
68). A steel reproduction of the original six-bladed design has been
installed on the *Great Britain*. In contrast to paddle steamships, it was
necessary to arrange the engines such that they could drive the propeller
shaft which ran along the longitudinal centre line of the ship and to
achieve this, the four cylinders working in pairs were inclined inwards
and upwards at an angle of 60 degrees (see Fig 69). Connected to the
overhead crankshaft was an 18ft diameter wheel which projected above
deck level. It was housed with skylights for viewing. The drive to the 6ft
diameter wheel on the propeller shaft below was effected by means of
toothed endless chains.

The engines were made by the Great Western Steamship Company
and when the vessel was completed in July 1843 the building committee
of Messrs Claxton, Guppy and Brunel claimed to have built the 'first

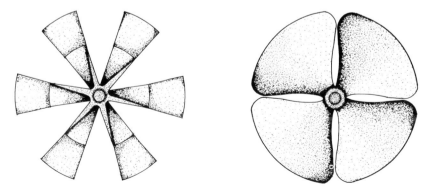

Fig 68 The six- and four-bladed propellers for the SS *Great Britain*

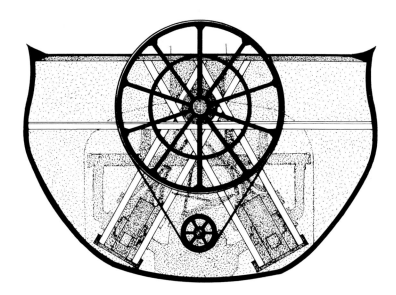

Fig 69 The four cylinders working in pairs were inclined at an angle of 60 degrees

modern ship' and the first to cross the Atlantic with screw propulsion. It has been emphasised that many of the design aspects of the *Great Britain* were not original, but Brunel combined these on a scale which brought about the step forward from ships of a few hundred tons to one with a displacement of over 3,500 tons in four years.

Of particular merit is the work on the screw propeller, and shortly after the decision to adopt screw propulsion on the *Great Britain*, the Admiralty requested a copy of possibly the most erudite of Brunel's reports, which is reproduced in full in I. Brunel's biography of his father. Paddle wheels were not suitable for warships and the Navy were interested in testing screw-propelled vessels. After numerous delays and ill-mannered behaviour to Brunel regarding his advice, an official test of screw propulsion was made but it was after the launch of the *Great Britain*. HMS *Rattler*, 888 tons, 200hp was fitted with a screw propeller and trials were carried out in 1843 and 1844, including a tug of war against HMS *Alecto*, a paddle ship. The outcome of this was that in 1845 the Admiralty ordered more than twenty vessels to be fitted for screw propulsion.

The early voyages of the *Great Britain* have been mentioned previously and it was on 22 September 1846 that she was put to her greatest structural test. With 180 passengers aboard she left Liverpool on her outward voyage to New York and the apparently normal course was to pass the Isle of Man to the south of the Calf of Man and then to turn north and pass between the Isle of Man and the north east coast of

Ireland. For reasons which have never been fully explained, the St Johns light at the northern tip of Dundrum Bay was mistaken for the Chicken Rock light off the Calf of Man and the ship ran ashore in the bay. On the following day, the passengers were landed at ebb tide. The ship was visited by Captain Claxton who ordered her to be driven by sail further onto the beach for safety. She was visited by Brunel in December 1846 and this prompted a letter to Captain Claxton:

Don't let me be understood as wishing to read a lecture to our Directors; but the result, whoever is to blame, is, at least in my opinion, that the finest ship in the world, in excellent condition, such that £4,000 or £5,000 would repair all the damage done, has been left, and is lying, like a useless saucepan kicking about on the most exposed shore that you could imagine, with no more effort or skill applied to protect the property than the said saucepan would have received on the beach at Brighton.   (I. K. B. 10 December 1846)

Brunel reported extensively on the damage and considered the ship perfect apart from the bottom being holed in several places. He proposed a method of protecting the ship by means of a skirt of faggots lashed together. It was Brunel's view that the ship should be raised by mechanical means and the leaks made good. In May 1847 a salvage expert, Alexander Brummer, devised a means of lifting her by levering up the bow and at the same time raising the forward half of the vessel by running ropes under the hull and over vertical posts. Boxes of sand were attached to the ropes which pulled the ship up. Eventually the ship was floated and towed to Liverpool. As a result of the efforts of Brunel, and later the SS *Great Britain* Project team, the ship has survived being holed twice, first at Dundrum Bay in 1846 and ninety-one years later at Sparrow Cove. Her presence in the Great Western dry dock at Bristol enriches the influence of Brunel on the city and she complements the Clifton Suspension Bridge and Temple Meads Station.

PSS *GREAT EASTERN*

A popular tourist attraction in London is the river trip from Tower Pier to Greenwich Pier to visit the National Maritime Museum and the *Cutty Sark*. The river launches pass through what is left of London's dockland, cross the line of Marc Isambard Brunel's Thames Tunnel at Rotherhithe, and a short distance from Greenwich, on the north bank at Millwall on the Isle of Dogs, pass a bold statement crudely painted on the concrete capping beam at the top of the sheet pile river wall – 'The Great Eastern was launched here'. Apart from a few relics buried deep in the Thames mud, and souvenirs from the breaking up in 1889, little remains of

Brunel's final contribution to naval architecture, the 27,380-ton PSS *Great Eastern*. Although the primary intention of this book is to chronicle the works of the Brunels that exist today it would be inappropriate to omit a description of Isambard Kingdom Brunel's contribution to the design of 'a great ship'.

In 1851, Brunel advised the directors of the Australian Royal Mail Company on the appropriate type of ship for the Australian mail run. Brunel's suggestion of a 6,000-ton ship was rejected and subsequently Brunel invited tenders for two ships with displacements slightly less than the *Great Britain*. Brunel accepted the tender prepared by John Scott Russell, a naval architect and ship builder, and this was the start of their ill-fated relationship, although they had met previously at various meetings of learned societies. The merits of the contributions of Brunel and Russell to the development of iron ships have been dealt with at length by L. T. C. Rolt (1957) and George S. Emmerson (1977). From the evidence presented, it is apparent that until the publication of Emmerson's biography, John Scott Russell's contribution to iron ship building had been greatly understated. The following brief description of his career up to the time of his association with Brunel on the *Great Eastern* project suggests his knowledge of ship design and construction was at least the equal of Brunel's.

John Scott Russell was born near Glasgow on 9 May 1808 and obtained his Master of Arts Degree in 1825. During the 1830s he carried out theoretical and experimental studies of the form and structural arrangement of ships, and was aware of the theoretical work of Young and Euler. A significant aspect of his research was the conception of the 'Wave line' hull form, but even more important was the structural design of a 120ft steamer, the *Storm*.

Transverse frames or ribs were not used and the hull was stiffened by a

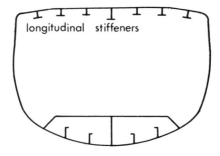

Fig 70 Simplified structural section through a modern steel ship with longitudinal stiffeners

number of transverse bulkheads and the longitudinal lap joints were reinforced with T bars. This design was very close to that adopted for modern ships in which longitudinal stiffeners are used (see Fig 70). The *Storm* was designed and built prior to the *Great Britain* and it could be argued that the structural conception was more advanced. By 1850, Russell had designed and built a number of iron ships which incorporated longitudinal framing and a partial iron upper deck. However, Brunel had built an iron ship of over 3,500-tons displacement which had survived being grounded and holed. Another structure of tubular form, relevant to the final design of the *Great Eastern* was the Britannia Tubular Bridge, completed in 1850 (see Introduction). A section through the tubes is shown in Fig 71. The top and bottom chords or flanges were of cellular construction and the vertical sides or webs, stiffened by vertical and longitudinal ribs. The main tubes were 138ft longer than the hull of the *Great Britain* and each weighed 1,600 tons. The combined knowledge of Brunel and Russell regarding the strength and construction of iron steamships, and their knowledge of the work of Stephenson and Fairbairn on the design of the Britannia tubes, augured well for the design of a ship of nearly eight times the displacement of the *Great Britain*.

Brunel's association with the Australian Royal Mail Company led to the appearance in his sketch books of a design for a ship of 700ft length and 60ft depth with the hull supported at two points 400ft apart. This

Fig 71 A cross section through the Britannia Tubular Bridge turned through 90 degrees

was followed by discussions with his colleagues, including Russell, and also with the newly formed Eastern Steam Navigation Company. Following these discussions, Brunel was appointed engineer to the company, and in May 1853 the board adopted his recommendations to accept the tenders of John Scott Russell for construction of the hull and paddle engines of the 692ft long ship and that of Messrs James Watt & Company for the screw engines. It had been calculated that the required power could only be achieved by a combination of paddles and screw, the contribution of the paddles being about sixty per cent of that of the single screw. The diameter of the paddle wheels was 53ft and the screw 24ft.

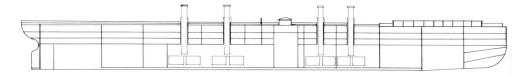

Fig 72 PSS *Great Eastern* in longitudinal and transverse section

The general structural arrangement of the ship in longitudinal and transverse section as shown in Fig 72 can be attributed to Brunel, and this was acknowledged by Russell. The ship was divided into watertight compartments by ten bulkheads, eight of which extended to the level of the upper deck. Up to the water mark, the hull was of cellular construction consisting of two skins of 1in plates spaced 2ft 10in apart, and between these ran longitudinal members at a spacing of 6ft. The upper deck was also of cellular construction and additional strength was provided by two 350ft long, 36ft deep longitudinal bulkheads. This immensely strong structure survived an 85ft hole being torn out of her hull during an Atlantic run in 1862. To Russell can be attributed the general lines of the ship, and the practical means of implementing Brunel's ideas. This was a monumental task involving the rolling of about 30,000 plates and the use of 3 million rivets. The plate size was generally 10ft by 2ft 9in and the clinker laid method used on the *Great*

Britain was superseded by the 'in and out' system (Emmerson) which involved much less cutting.

During the slow progress of the construction of the hull which, because of its great length, was built parallel to the river bank for sideways launching, the predictable financial problems arose and the relationship between Brunel and Russell was not improved by the fact that, as engineer for the project, Brunel was reponsible for authorising payments to Russell. Eventually the hull was completed and the launch was planned for 3 November 1857. Prior to this, details of the launch procedure had been carefully planned by Brunel and it was decided to adopt iron to iron sliding surfaces. Two inclined planes (1 in 12) were constructed under the ship, each 120ft wide extending a distance of over 300ft down to the low-water mark, were supported on a network of timber baulks resting on a concrete bed with further support obtained from a grid of piles. The problem facing Brunel was to slide a load of about 12,000 tons about 250ft. After a number of attempts and the assistance of hydraulic rams, the ship took to the water on 31 January 1858. Further funds were required to complete the ship and a new company was formed to complete the fitting out.

The steam dredger *Bertha* at the Exeter Maritime Museum (*Derrick Beckett*)

At about this time, the original name, the *Leviathan* was changed to the *Great Eastern*. Towards the end of 1858, Brunel's health was failing. He was suffering from a severe kidney infection, nephritis, and his doctors recommended a winter in a warmer climate. Brunel returned the following May to immerse himself in the final problems leading up to the first sailing of the *Great Eastern*. His last visit to the ship was on 5 September 1859 and around midday he suffered a stroke. Ten days later, shortly after receiving news of the explosion of one of the feed heaters at the base of a funnel whilst the *Great Eastern* was off Portland in the English Channel, Brunel died at his Duke Street home in London. Following Brunel's death, the *Great Eastern* spent her most productive years laying submarine cables from Europe to America and Bombay to Aden. She ended her days in a similar undignified manner to more recent queens of the Atlantic, as a show boat on the River Mersey.

BERTHA

To end this chapter on a brighter note, the oldest operational steam driven vessel in England, the dredger *Bertha*, can be seen on the River Exe at Exeter Maritime Museum, Devon. Although there is no documentary evidence, it is believed to have been designed by Isambard Kingdom Brunel and was built in 1844. She was probably built in Bristol but assembled in Bridgwater where she had the task of clearing the mud from the dock. A large blade was fitted on the bottom of the hole in the stern. The blade was lowered into the mud and the boat would then haul itself along dragging the mud with it as it did so, until finally it dragged it out into the stream of the river, where the river itself completed the task. She is in steam annually.

GAZETTEER

London

London is the most convenient base from which to commence a tour of the thirty-five principal locations. Most visitors will have a limited amount of time available and thus it is worthwhile to spend some time on planning the sequence of visits to suit individual requirements.

In many cases they can be combined with visits to other places of interest not connected with the works of the Brunels. With the exception of Hacqueville (35), it is possible to visit all the other locations individually without an overnight stop, but in some instances this necessitates spending most of the day in a train or car. The train would appear to be the obvious form of transport, but as indicated previously, peering out of a window of a High Speed Train is not generally the best. means of observing the stations and other constructions on the route. Further, the timetables are not always convenient. The author has visited all the locations outside London by train and car and the more appropriate mode of transport will be given as each location is described. Directions are given, where appropriate, from the nearest motorway junction and the nearest railway station is also listed. All directions are related to visitors travelling from London. The following references are useful in planning a journey:

1 *London Street Atlas & Index A-Z* (Locations 1 to 12).
2 *Nicholsons Guide to the Thames* (Locations 9, 10, 11, 13, 14, 17 and 18).
3 Banks, F. R., *The Penguin Guide to London* (Locations 1 to 12).
4 The following Ordnance Survey Sheets (1 : 50000, approx $1\frac{1}{4}$in to 1 mile) cover the rail route from Paddington to Penzance, via Bristol.

Sheet 176   West London
Sheet 175   Reading & Windsor
Sheet 174   Newbury & Wantage

Sheet 173   Swindon & Devizes
Sheet 172   Bristol & Bath
Sheet 182   Weston-super-mare & Bridgwater
Sheet 192   Exeter & Sidmouth
Sheet 202   Torbay & South Dartmoor
Sheet 201   Plymouth & Launceston
Sheet 200   Newquay & Bodmin
Sheet 204   Truro & Falmouth
Sheet 203   Lands' End & the Lizard.

Details of all maps published by the Ordnance Survey are available from the Ordnance Survey, Romsey Road, Maybush, Southampton SO9 4DH.

5 Bristol City Centre A-Z Street Map – Geographers A-Z Map Co. Ltd.

1 *Paddington Station and Great Western Royal Hotel*
(London Underground: District, Circle, Metropolitan and Bakerloo lines)
The magnificence of Brunel's arch roof is now complemented by the High Speed Trains which serve Bristol, Cardiff, Swansea, Exeter, Plymouth and Penzance. The roof is best observed by walking down Platform 1 (there is no ticket barrier) past the directors' balcony, which faces the transept, as far as the footbridge which crosses the platforms and provides access to the Underground. From the footbridge, it is possible to obtain a close view of the wrought-iron arch ribs and the decorative work in cast iron which is bolted to them. There is direct access to the four-star Great Western Royal Hotel (01 723 8064) from the station concourse and the corridor leads into the rear of the Rendezvous Lounge. In front of the Rendezvous Lounge is the reception, to the right of which is the new Brunel bar and restaurant designed in a Victorian theme.

2 *Kensal Green Cemetery*
(Kensal Green, British Rail Midland Region and London Underground)
Kensal Green Cemetery, Harrow Road (01 969 0152) is sandwiched between the main lines from Euston and Paddington. The Brunel family grave is located about half way along the central path which leads from the entrance to the church, on the left hand side. It is not easy to find but directions can be obtained from the attendants at the entrance. Opening hours: Monday to Saturday 9.00 to 16.00 (18.00 summer), Sunday afternoon.

3 *Cheyne Walk*
(Nearest Underground South Kensington, Bus Routes 19, 45, 49 to Battersea Bridge)
Lindsey House retains its original façade and was the home of Marc Isambard Brunel between 1811 and 1826. The house is situated in Cheyne Walk to the west of Battersea Bridge and is about fifteen minutes walk from South Kensington Station. Take the 19 or 45 bus from the station to Battersea Bridge. Lindsey House is a National Trust property but there is no admission.

4 *Embankment*
(Nearest Underground, Temple)
The statue of Isambard Kingdom Brunel is at the junction of Temple Place and the Victoria Embankment almost opposite HMS *Discovery*. Brunel gazes upstream and his view of the red brick piers which are all that remains of Hungerford Suspension Bridge (now Charing Cross Railway Bridge) is almost blocked by Waterloo Bridge (1937–45), one of the most distinguished reinforced concrete bridges built in England.

5 *Westminster Abbey*
(Nearest Underground, Westminster)
There is a memorial window to Isambard Kingdom Brunel in Westminster Abbey. Entering the nave through the west entrance, the window faces the Deanery between the second and third piers which separate the nave from the south aisle. Admission to the nave, daily from 8.00 to 18.00. On certain public holidays the Abbey is open for services only.

6 *National Gallery*
(Nearest Underground, Charing Cross or Leicester Square)
The Gallery holds one of the most extensive collections of paintings in the world, including Joseph Mallord William Turner's 'Rain Steam & Speed – the Great Western Railway', canvas 35.75in by 40in. It is to be found amongst a collection of paintings from the British School (room 34). Admission free, Weekdays 10.00 to 18.00, Sundays 14.00 to 18.00. Closed certain bank holidays. A postcard and large scale reproductions of Turner's painting are available.

7 *National Portrait Gallery*
(Nearest Underground, Charing Cross or Leicester Square)
In Room 14, there is a portrait of Marc Isambard Brunel (p. 21) painted

by Samuel Drummond in 1836 with the Thames Tunnel in the background. The portrait of Isambard Kingdom Brunel (p. 26) by J. C. Horsley, 1857, is not on display. Postcard reproductions are available together with R. Howlett's classic photograph of Isambard Kingdom Brunel (p. 13).

8 *Science Museum*
(Nearest Underground, South Kensington)
It is necessary to be extremely single minded not to be diverted by the numerous stimulating exhibits at the Science Museum, but a few hours are required for those relevant to the Brunels alone. Starting with the ground floor, head for the exhibits on rail transport, rooms 7 and 8. There is a section on permanent way with Great Western rail sections used from 1838 onwards. In room 9A (Roads, Bridges & Tunnels) there is a sectional model of the Thames Tunnel, showing the Rotherhithe shaft and the tunnelling machine. On the first floor, room 21 (Hand and Machine Tools), there are models of the Portsmouth block-making machinery, the earliest large-scale use of the machine tool for mass production and a description of the stages in the manufacture of pulley blocks. Room 49 on the second floor contains models of the *Savannah*, *Sirius*, *Great Western*, *Great Britain* and *Great Eastern*. There are also sectional models of the *Great Eastern*'s paddle engines and the *Great Britain* engines. Hours of opening: weekdays including Saturdays and Bank Holidays 10.00 to 18.00, Sundays 14.30 to 18.00, Closed Good Friday, Christmas Eve and Day, Boxing Day and New Year's Day.

9 *Wapping*
(East London section of the London Underground)
It is worth a ten minute stop at Wapping, on route to Rotherhithe, to observe the twin horseshoe arches of the tunnel as they dip under the Thames at the south end of the platform. The lifts ascend to ground level within the original shaft and in the entrance to the station, there is a commemorative plaque to the Brunels (see Chapter 10) erected by London Transport in 1959.

10 *Rotherhithe*
(East London section of the London Underground)
Rotherhithe Station is a short distance to the south of the Thames Tunnel shaft and the station entrance faces the approach to Rotherhithe road tunnel, completed in 1908. On leaving the station entrance, walk back towards the river along Railway Avenue and the tunnel shaft and

restored engine house are to be seen on the left. Nearby are the Mayflower public house with a fine view across the river to Wapping and a number of restored dockland buildings.

11 *Greenwich: The National Maritime Museum*
(British Rail Southern Region, Maze Hill)
The train journey from Charing Cross, Waterloo or London Bridge to Maze Hill takes about twenty minutes. In the summer, an alternative route is to take a river bus from the Tower of London to Greenwich Pier. The bus passes the site of the launching of the *Great Eastern* and passengers alight at Greenwich Pier adjacent to the permanent berth of the *Cutty Sark*. Proceed along William Walk past the Royal Navy College to the western entrance of the Museum Grounds. The Museum can justifiably claim to be one of the most beautiful in Britain and again single mindedness is necessary to keep to exhibits relevant to the Brunels, which are to be found in the east wing. The east wing is close to the principal entrance to the main buildings and is at the south-eastern corner at the end of Park Vista which leads down from Maze Hill Station. There are galleries devoted to the development of wooden ships, steamships and ship structure.

There is a superb model of the *Great Britain*, and amongst the paintings, is 'Building the Great Eastern' by William Parnatt (1813–69). Several prints of Brunel's steamships are available from the bookshop at the entrance to the east wing after paintings by J. Walter, including the launching and trials off Lundy. Open from 10.00 to 17.00 Monday–Saturday and 14.30 to 18.00 on Sunday. Admission to the Museum is free.

Locations Outside London

12 *Wharncliffe Viaduct Hanwell*
(Hanwell and Elthorne, British Rail Western Region)
One of Brunel's first and most impressive brick structures located about seven miles west of Paddington Station. At junction 1 of the M4 motorway, turn onto the A406 (Gunnersbury Avenue) travelling north for about $1\frac{1}{2}$ miles and then turn left onto the A4020 (Uxbridge Road). The A4020 crosses the River Brent about two miles west of the junction with the A406 and the viaduct can be seen on the right straddling the recreation grounds. Parking on the A4020 is difficult so turn right at the junction of the lower Boston Road (A3002) and the A4020 into Half Acre Road and park at the top adjacent to the recreation ground. The

suburban train service from Paddington to Hanwell (15 minutes) runs at approximately 30 minute intervals and the viaduct is a short distance to the west of the station. Traffic congestion in central and outer London is generally such that the train is the preferable mode of transport.

13 *Thames Bridge Windsor*

(Windsor Central, British Rail Western Region)
A wrought-iron bowstring girder bridge on the Slough to Windsor spur from the main line. Take the High Speed Train (14 minutes) or the suburban service (45 minutes) from Paddington to Slough. There is a connecting service from Slough to Windsor Central at approximately 20 minute intervals and the bridge is approached via a long brick viaduct. Windsor Central Station is a few minutes walk from the bridge and it is of course located close to another well-known place of interest. Travellers by car, leave the M4 motorway at junction 6 and take the Windsor spur which crosses the Thames about one mile south of the motorway. At the junction with the A308, turn left into Windsor and left again at the first set of traffic lights. There is a car park adjacent to Windsor Lido and access to the bridge is via the pleasure grounds which back on to the car park.

14 *Thames Bridge Maidenhead*

(Maidenhead, Taplow, British Rail Western Region)
The elegant bridge at Maidenhead with twin semi-elliptical arches is best viewed from the towpath and is located a good mile from both Taplow and Maidenhead stations. Further, the train service to both stations is infrequent. It is more convenient to use a car and it is suggested that the visit could include Sonning Cutting and Mortimer Station. Leave the M4 motorway at junction 7 and turn left into Maidenhead (A4) at junction 7A. At the approaches to the road bridge carrying the A4 over the Thames, turn left again and the rail bridge is about $\frac{1}{4}$ mile downstream. Parking is possible in an adjacent side road.

15 *Sonning Cutting*

(Reading, British Rail Western Region)
Reading Station is not within easy walking distance of Sonning Cutting. Motorists should leave the M4 motorway at junction 10 and take the A329(M) spur to Reading. At the junction with the A4 turn right onto the A4 which crosses Sonning Cutting about one mile east of the junction. There are twin bridges at the crossing, the west bound carriageway of which is supported by Brunel's three span brick arch.

Due to parking difficulties, it is preferable to continue to the next roundabout and then turn right onto a minor road. Follow this road for a short distance and take the right-hand fork which crosses the cutting about $\frac{1}{4}$ mile south of the A4. The bridge is a good vantage point for observing the extent of the cutting and the High Speed Trains. Looking west, note the metal arch bridge which replaced Brunel's original five span timber bridge. At this point the cutting is about 60ft deep.

16 *Mortimer Station*
(British Rail, Southern Region)
Mortimer Station with its many Brunelian characteristics is now under the control of British Rail Southern Region and is located on the Reading to Basingstoke line about 6 miles south of Reading. The fastest train service from Paddington to Reading is 27 minutes, by High Speed Train. The connecting service to Mortimer is at approximately hourly intervals, journey time 9 minutes. Motorists should leave the M4 at junction 11 turning right onto the A33 Reading to Basingstoke road. At Three Mile Cross, about half a mile south of junction 11, turn right onto a minor road which passes through Grazeley and crosses the railway about $1\frac{1}{2}$ miles north of Mortimer Station.

17 *Thames Bridge Basildon*
(Goring and Streatley, British Rail Western Region)
Basildon and Moulsford bridges are two further examples of Brunel's use of brick arches for a major river crossing and are located on a picturesque stretch of the river Thames about one mile south and two miles north of Goring respectively, which is set in a heavily wooded valley. Leave the M4 motorway at junction 12 and take the A340 to Pangbourne. Pass under the railway bridge with Pangbourne Station on the left and follow the A329 to the junction with the B4009. Turn right, cross the Thames, and after passing through the village turn right again at the railway bridge. Continue a short distance past the station and take the right-hand fork to Gatehampton farm. The land adjacent to the bridge is private and permission is required for access. Alternatively, the bridge can be reached via the towpath from Goring (east bank), about one mile. The train service from Paddington to Goring and Streatley is generally at hourly intervals, journey time about 75 minutes.

18 *Thames Bridge Moulsford*
(Goring and Streatley, British Rail Western Region)
The same route as for Basildon Bridge but turn left at the railway bridge

at Goring and follow the B4009 to South Stoke. Turn into the village passing under the railway bridge and follow the road through until it reaches the river (Beetle and Wedge Hotel on opposite bank but no ferry). Access to the bridge is via the towpath on the east bank, about $\frac{1}{2}$ mile. Allow for walking a distance of about 7 miles if both bridges are to be visited in one day with Goring as the starting point.

19 *Culham Station*
(British Rail, Western Region)
Few trains on the Paddington to Oxford line stop at Culham and the journey by car could embrace Basildon, Moulsford, and Steventon. Again take the same route as for Basildon but continue on the A329 through Streatley and Moulsford. About one mile north of Moulsford the road crosses the railway and the hotel for Brunel's Wallingford Road station can be seen on the right. Continue through Wallingford until the A423 is reached. Turn left onto the A423 and continue through Dorchester travelling towards Oxford. Then take the left fork (A415) Dorchester to Abingdon Road and about 3 miles further, Culham Station is on the right.

20 *Steventon*
(Didcot, British Rail Western Region)
Steventon station buildings have now been demolished but the district superintendent's house with Brunelian characteristics remains. Didcot Station is several miles walk from Steventon and the only practical means of access is by car. Culham Station is a few minutes' drive from Steventon, so follow the previous directions to Culham Station and continue along the A415 to Abingdon and turn left onto the A415 to Abingdon and left again onto the A4017. Continue about five miles southwards on the A4017 passing through Drayton and the house is in the station yard on the right-hand side of the road as it approaches the bridge crossing the railway.

21 *Swindon*
(British Rail, Western Region)
There is a frequent High Speed Train service from Paddington to Swindon, the journey time being about one hour. For access by car, leave the M4 motorway at junction 15 and head towards Swindon. At the next roundabout, bear left onto the A345 and follow the signposts to the town centre where there is ample parking close to the new Brunel Shopping Centre. The Great Western Railway Museum and the adjacent railway

village are located in Faringdon Road which is a few minutes' walk from the station. In addition to the replica of the 2–2–2 *North Star*, the museum contains many other items of interest relating to Brunel and the Great Western Railway. Open weekdays 10.00 to 17.00, Sundays 14.00 to 17.00. Closed Good Friday and Christmas. The railway village is a stimulating example of the benefits of conservation and together with the museum merits a full day's visit.

22 *Chippenham*
(British Rail, Western Region)
The scheduled journey times from Paddington to Chippenham vary from just under the hour to 1 hour 17 minutes, depending on the number of intermediate stops. For access by car leave the M4 motorway at junction 17 turning left onto the A429 which leads into the centre of Chippenham. The massive arch and viaduct are at the junction of the A4 and A 420, a short distance west of the railway station.

23 *Box Tunnel*
There are no intermediate stations between Chippenham and Bath but the A4 crosses the railway a short distance from the west portal of the tunnel. Follow the route to Chippenham and then take the A4 travelling west towards Bath. The road crosses the railway about 7 miles west of Chippenham but parking close to the bridge is dangerous, but it is possible in Box village. It is well worth waiting at the bridge to watch the High Speed Train emerge from the west portal.

24 *Bath*
(British Rail, Western Region)
The route chosen by Brunel for the section of the line between Box Tunnel and Bath Station offers some impressive views of this elegant Georgian city and the surrounding countryside. A full day's visit is essential and the 1 hour 10 minute journey from Paddington by High Speed Train is thoroughly recommended, the last few minutes of which are an absolute pleasure. Bath's celebrated Roman and Georgian architecture is complemented by Brunel's railway architecture, the retaining walls and bridges in Sydney Gardens, St James Bridge to the east of the station and the long viaduct to the west of the station. Sydney Gardens are about 15 minutes walk from the station. For visitors who must travel by car, leave the M4 motorway at junction 18, turning left onto the A46 and proceed into Bath, about 8 miles.

25 *Bristol*

At least a full day is required to visit Bristol Temple Meads Station, the SS *Great Britain*, and Clifton Suspension Bridge. There is a frequent High Speed Train service from Paddington and the journey time is about 90 minutes. Travellers by car along the M4 motorway have a number of options available at the approaches to Bristol. The author has experienced the minimum congestion by joining the M5 (westbound) at junction 20 of the M4 and then turning left off the M5 at junction 17 and taking the A4018 through Westbury on Trym and following the signposts to Clifton and the Suspension Bridge. Thus it is more convenient for travellers by car to visit the three locations in the reverse order. The A–Z street map of Bristol is essential and directions are given starting from Bristol Temple Meads Station. The original station can be reached by walking across to the right of the forecourt from the booking hall, where a plaque indicates the date of the opening, at the entrance to what is now a car park.

26 *SS Great Britain*

The SS *Great Britain* is about 20 minutes walk from Bristol Temple Meads Station. Proceed via Clarence Road and Commercial Road into Cumberland Road. Follow the River Avon (new cut) and turn right into Gas Ferry Road. Great Western Dock is at the end of Gas Ferry Road and the opening hours are 10.00 to 17.00 winter and 10.00 to 18.00 summer. For further details, contact SS *Great Britain*, Great Western Dock, Gas Ferry Road, Bristol BS1 6TY. Tel: Bristol (0272) 20680. Car parking and refreshments are available.

27 *Clifton Suspension Bridge*

It is possible to walk from the *Great Britain* to Clifton in about half an hour. On leaving Gas Ferry Road, turn right and proceed via Cumberland Road, Avon Crescent and Merchants Road to Hotwells Road. Turn left into Hotwells Road and there then follows a steep climb via Hopechapel Hill, Granby Hill and Sian Hill up to the bridge. The effort will be rewarded by spectacular views of the Avon Gorge and Bristol from the suspension bridge.

28 *Bleadon*

(Weston-super-Mare, British Rail Western Region)
The 'flying' arch at Bleadon is about 2 miles south of Weston-super-Mare Station. Leave the M5 motorway at junction 21 turning right onto the

A370. Follow the A370 through Weston-super-Mare until it crosses the railway line on the southern edge of the town. The 'flying' arch bridge can be seen looking north from the railway bridge. Looking south, the line continues dead straight across Bleadon Level.

29 *Exeter*
(Exeter St Davids, British Rail Western Region)
Brunel's dredger *Bertha* and many other interesting aspects of maritime archaeology can be seen at Exeter Maritime Museum. By road, follow the M4/M5 motorway route and leave the M5 at junction 30. The museum is on the river and motorists should follow the Maritime Museum signs as they approach the centre of the city. The museum is open every day of the year except Christmas Day, Winter 10.00 to 17.00 (October to May), Summer 10.00 to 18.00 (June to September). For further details contact The Director, ISCA Ltd, The Quay, Exeter, Devon. Tel: Exeter 58075. The fastest journey time from Paddington to Exeter St Davids by High Speed Train is 2 hours 29 minutes.

30 *Starcross*
(British Rail, Western Region)
Between Exeter and Teignmouth, the railway line follows the west bank of the River Exe and the coastline, passing through Starcross and Dawlish. This is one of the most interesting sections of the route from Paddington to Penzance but unfortunately very few trains stop at Starcross and there is no Sunday service in winter. By car, follow the M4/M5 motorway route to Exeter. Leave the motorway at junction 30 and travelling into Exeter turn left onto the westbound A38. After crossing the River Exe and Exeter Canal turn left onto the A379 which leads to Starcross. Brunel's engine house, built in the style of an Italian bell tower, is sandwiched between the A379 and the railway at the southern end of Starcross Station.

31 *The Royal Albert Bridge Saltash*
(Plymouth, British Rail Western Region)
Although there is a station at Saltash, the service is so infrequent that it is preferable to travel by Inter-City from Paddington to Plymouth, journey time about $3\frac{1}{2}$ hours. There is a frequent bus service from the entrance to Plymouth Station and passengers should alight after crossing the Tamar Bridge. Walk down the hill to the river level passing under the approach viaduct to the main spans to obtain a real impression of the scale of the structure. It is also possible to walk across the Tamar

Suspension Bridge to inspect more closely the arch/chain deck structure of the Royal Albert Bridge which is adjacent. The round trip from Paddington to Saltash is possible in one day but over 7 hours will be spent on the train. A one day journey by car is not recommended.

32 *Liskeard Station*
(British Rail, Western Region)
Between Newton Abbot and Penzance there are a number of remains of the masonry piers to Brunel's timber viaducts and Liskeard is chosen as most convenient, as there are two major viaducts immediately to the east (Liskeard) and the west (Moorswater) of the station. The road crossing the railway is supported by a 'flying' arch bridge. The round trip by train from Paddington is about 13 hours, which allows over 4 hours to inspect the remains of the original viaduct piers.

33 *Chatham Dockyard*
(Chatham, British Rail Southern Region)
Marc Brunel's sawmill, boiler house and chimney form part of an impressive collection of dockyard buildings at HM Naval Base, Chatham. Permission is required to visit these buildings and should be obtained from the Office of the Port Admiral, HM Naval Base, Chatham, Kent. The 34-mile car journey from central London is tedious but there is a frequent train service, about 40 minutes from Victoria. The entrance to the dockyard (Pembroke Gate) is about two miles from the station and there is a frequent bus service.

34 *Portsmouth*
(Portsmouth Harbour, British Rail Southern Region)
Isambard Kingdom Brunel was born in Britain Street, Portsea on 9 April 1806. The house has now been demolished but there is a commemorative plaque. The site of the house is a few minutes' walk from HMS *Victory* and the ship is a fine example of naval architecture prior to innovations such as diagonal bracing. The Portsmouth block-making machinery, designed by Marc Isambard Brunel, was used for the mass production of pulley blocks, some of which were no doubt used in refits of HMS *Victory* (Tel: 0705 815615). Opening hours for HMS *Victory* are Monday to Saturday 10.30 to 17.00 (winter), 10.30 to 17.30 (summer), Sundays 13.00 to 16.00 (winter), 13.00 to 17.00 (summer). There is a frequent train service from Waterloo to Portsmouth Harbour.

35 *Hacqueville*

(SNCF Gisors, Route 306)

For the enthusiast, a pilgrimage to 'La Ferme Brunel' at Hacqueville is possible by the following rail/sea route. The first stage of the journey is from London (Victoria) to Dieppe Maritime via Newhaven. From Dieppe (Gare), Gisors is about 1 hour 20 minutes by train (route 306 Dieppe-Paris). There is a privately operated bus service from Gisors to Etrepagny which is 2½ miles from Hacqueville. For full details of train times etc, contact French Railways (SNCF), French Railway House, 179 Piccadilly, London W1V 0BA (01 493 4451/2).

TECHNICAL APPENDICES

Early Developments in Structural Mechanics

Consider a vertical bar of cross-sectional area A rigidly fixed at one end and free at the other (Fig 73). The bar is subjected to a load N which produces extension of the material fibres (tension) and thus the intensity of load per unit area (ignoring the self-weight of the bar) is given by

$$f = \frac{N}{A} \qquad \text{(equation 1)}$$

where, in engineering terms, f represents stress. At a load of, say, N_r the bar will fail in tension due to rupture of the material fibres and thus the intensity of load per unit area or rupture stress f_r is

$$f_r = \frac{N_r}{A}$$

$$N_r = f_r \times A$$

If we consider the bar is subjected to its self-weight only then its length can be extended until its self-weight reaches the rupture load N_r. If   represents the weight per unit volume of the bar and L_r is the length of the bar at which rupture occurs (Fig 74) then

$$\text{weight of bar} = \gamma \times A \times L_r$$

$$\text{thus} \qquad N_r = \gamma \times A \times L_r = f_r \times A$$

$$\text{and} \qquad L_r = \frac{f_r}{\gamma} \qquad \text{(equation 2)}$$

This equation is a useful means of comparing the structural potential of various materials under a tensile force action and indicates that it is not possible to go on making structural members bigger and bigger as they will eventually fail under their own weight. The value of L_r for materials used by Brunel is established in Table 10. Masonry, brick and concrete are grouped as having very approximately the same low tensile strength and unit weight.

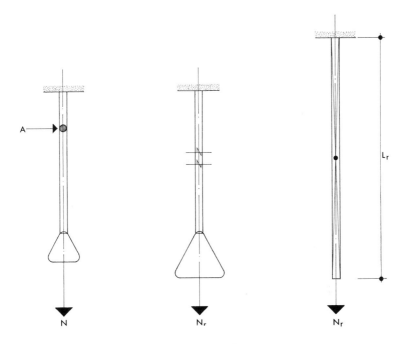

Fig 73 A bar subjected to a load which induces tension
Fig 74 L_r is the length of the bar at which rupture occurs

Table 10 gives, at first sight, the surprising result that timber has the best strength to weight ratio in terms of a tensile force action. However, nature has a few hundred million years start over man in developing a light weight material of cellular form (see Fig 75) a simplified model of which is a bundle of drinking straws glued together. As we have seen, Brunel used timber brilliantly for his viaducts in south Devon and Cornwall and was obviously aware that its extreme variability and the presence of natural defects such as knots and splits meant that high margins of safety must be introduced to limit stresses levels induced when the structures sustained their maximum loading.

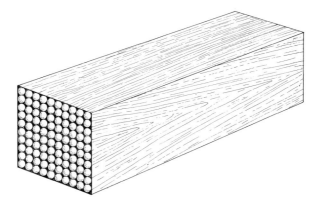

Fig 75 A simplified structural model of timber

MATERIAL	TIMBER (clear material)	WROUGHT IRON	CAST IRON	MASONRY/BRICK/ CONCRETE
Rupture stress in tension kN/m², f_r	40,000	350,000	110,000	3,000
Unit weight kN/m³, γ	5	76	76	24
$L_r = \dfrac{f_r}{\gamma}$ metres	8,000	4,605	1,447	125
$L_r = \dfrac{f_r}{\gamma}$ miles	4.92	2.83	0.89	0.077

Table 10 An approximate comparison of limiting lengths for various construction materials.

So far the performance of materials under a tensile force action only has been considered. Galileo also investigated the effect of bending on structural elements. Referring to Fig 76 in which a beam AB of length L is built into a wall at one end and at the free end is subjected to a load W the bending effect M, referred to as a bending moment, at the end A is given by

$$M = W \times L + \omega \times \frac{L}{2} \qquad \text{(equation 3)}$$

where ω represents the self-weight of the beam, the line of action of which is at a distance $\frac{L}{2}$ from A. Under the action of this bending moment Galileo assumed that all the material fibres were extended, that is, subjected to a tensile stress.

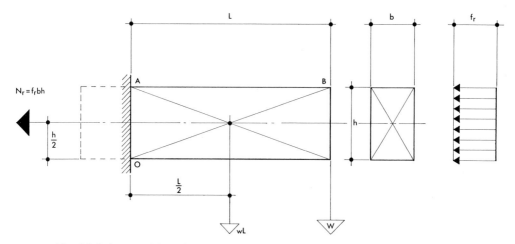

Fig 76 A beam of length L, built in at one end and free at the other

For the beam dimensions shown the total tensile force developed at rupture will be

$$N_r = f_r \, b.h$$

It was assumed that the section failed by rotation about 0 and thus at the point of failure we have the two moments in equilibrium, that is

$$N_r . \frac{h}{2} = M = W.L + \omega . \frac{L}{2}$$

now $N_r = f_r \, b.h$

thus $M = \dfrac{f_r}{2} \, b \, h^2$ (equation 4)

where M represents the applied bending moment and $\dfrac{f_r}{2} \, b \, h^2$ the internal resistance of the beam. This was the first mathematical expression for the strength of a beam in bending and it should be noted that for a given rupture strength the resistance of a beam to bending is proportional to its width times the square of its depth. Galileo was in error in assuming the tensile stress was uniform over the whole depth of the section as little more than a casual inspection of the deformed shape of the beam under the load W indicates extension will occur in the top fibres and compression in the bottom fibres, see Fig 77. This leads to an expression giving

$$M = \frac{f_r}{6} \, b \, h^2$$ (equation 5)

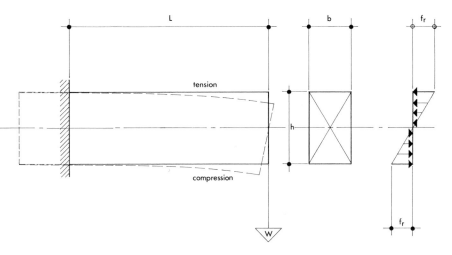

Fig 77 Beam deforms under load W, tension in top fibres and compression in bottom fibres

It is assumed that the maximum stress in tension and compression occurs at the top and bottom fibres respectively and reduces linearly to a neutral line at which the fibres are neither extended nor compressed. This equation represents the maximum strength of the section in bending and the concept of a neutral line was not appreciated by many engineers until the nineteenth century. If we consider a rectangular beam of span L and loading W then from equations 3 and 5 it can be shown that

$$W L = K_B b h^2$$

$$\text{or} \quad W = K_B \cdot \frac{b h^2}{L} \qquad \text{(equation 6)}$$

where K_B is a coefficient which is related to the rupture strength of the material and the end support conditions. There is no doubt that Brunel used an equation of this form which states in mathematical terms that the load a beam can carry is proportional to its width times depth squared and inversely proportional to its length.

In the simplest terms a beam of long span with small cross sectional dimensions can sustain only a small load. The influence of changing the shape of the cross section from the rectangular form will be considered later.

In Chapter 5, Fig 25 illustrates some tests carried out by Brunel on the resistance of square sections of timber columns to a compressive force. In the eighteenth century, Leonard Euler (1707–83) derived an equation for the ulitmate load of columns or struts which for square sections of side h and length L could be expressed in the form

$$N_E = K_c \cdot \frac{h^4}{L^2} \qquad \text{(equation 7)}$$

The coefficient K_c depends on an elastic property of the material and the end support conditions. Thus for a given cross-sectional dimension and end support condition, the ultimate compressive load is inversely proportional to the square of the length. Equation 7 is a mathematical expression which states that long slender columns buckle under a nominal load, see Fig 78. It is only when the column is extremely short that there will be compressive failure of the material fibres.

Finally, in this brief excursion into structural mechanics it is necessary to consider further the relationship between load and deformation. In the seventeenth century, Robert Hooke (1635–1703) published an anagram 'ceiiinosssttuv' which when solved reads 'ut tensio sic vis', the Latin for 'as the extension so is the force'.

To develop this, consider a bar loaded in tension (Fig 79) then for a load N_1 the bar will be extended by an amount x_1 and for a load N_2 by an amount x_2 and so on. For the load N_1 the change in length is x_1 and the ratio of the change in length to the original length is $e_1 = x_1/L$ and is termed a strain. At the load N_1, the stress is $f_1 = N_1/A$ and at the load N_2, the stress is $f_2 = N_2/A$. The relationship between stress (intensity of load per unit area) and strain (change

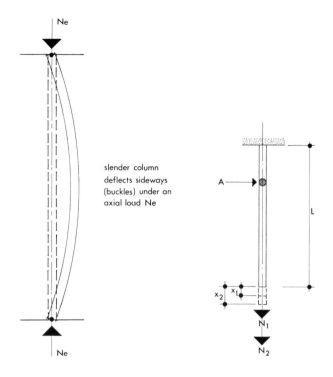

Fig 78 Buckling of a slender column
Fig 79 Extension of a bar due to loads N_1 and N_2

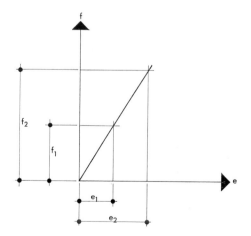

Fig 80 Relationship between stress and strain

in length divided by original length) is shown in Fig 80. The ratio stress divided by strain is termed modulus of elasticity that is

$$\text{Modulus of elasticity} = E = \frac{\text{stress}}{\text{strain}} = e \qquad \text{(equation 8)}$$

A material is considered as behaving in an elastic manner if there is a straight line relationship between load and deformation or stress and strain, that is, it obeys Hooke's law. The notion of modulus of elasticity was introduced by Thomas Young (1773–1829) at the beginning of the nineteenth century. It is a material property which relates resistance to deformation. The E values for timber, concrete and wrought iron are roughly in the ratio 1:2.5:20 and thus in terms of resistance to deformation, iron is a much superior material. It is necessary to consider materials in terms of their strength and resistance to deformation and for given material properties (ultimate stresses and modulus of elasticity), the designer can set about choosing the most appropriate form to resist a given force action (s). Observation of natural forms gives a useful guide to the influence of shape on strength, for example, the cellular cross section of a tree suggests the tube might be appropriate for both compression and tension. Indeed Brunel made extensive use of tubular members for long spans where need to keep the self-weight of the structure to a minimum was critical.

Gradients

In Chapter 2 reference was made to the force required to pull a wagon on a level track. The force N_F required to overcome friction (Fig 81) can be expressed as a function of the normal force, that is, the weight of the wagon W giving

$$N_F = \mu\, W \qquad \text{(equation 9)}$$

where μ is the coefficient of friction

If the wagon enters a gradient of an angle θ, (Fig 81), the normal force becomes $W \cos\theta$ and the component of W acting down the gradient is $W \sin\theta$. Thus the total force N required to pull the wagon up the gradient is given by

$$N = \mu\, W \cos\theta + W \sin\theta \qquad \text{(equation 10)}$$

For a gradient of 1 in 100, $\cos\theta$ approximates to unity and $\sin\theta$ to 0.01, thus

$$N = W + 0.01\ W$$

The value of μ W considered in Chapter 2 was 8lb (0.036kN) and the wagon weight W one ton (10kN). Thus

$$N = 8 + 0.01 \times 2,240$$

$$= 8 + 22.4$$

$$= 30.4\text{lb} \ (0.136\text{kN})$$

This is a simplified treatment of the problem of friction but explains why Brunel adopted shallow gradients at a time when the power of locomotives was limited. Applying equation 10 to the Great Western Railway with a gradient of 1 in 660 gives

$$N = 8 + 3.39$$

$$= 11.39lb \ (0.05kN)$$

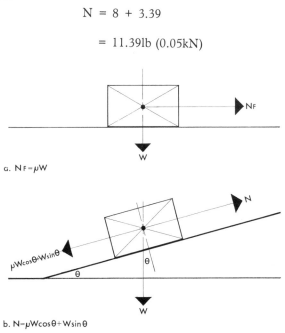

a. $N_F = \mu W$

b. $N = \mu W\cos\theta + W\sin\theta$

Fig 81 Force required to overcome friction and pull a train up a gradient

Design of Track

Early investigations of the strength of rails were carried out by considering them as simple beams on two supports. Approximate dimensions for an early plate rail are shown in Fig 82. The rail spans 3ft (0.92m) between stone blocks and for a static wheel load of 1 ton (10kN) at midspan, the maximum bending moment is 2.25 tons in. This results in extreme fibre stresses in the rail of about 3.0T/in² (46N/mm²) compression and 1.0T/in² (15N/mm²) tension. The gross weight of a wagon was unlikely to exceed 4 tons (40kN) and thus a maximum wheel load of about one ton (10kN) is appropriate. The maximum tensile stress of 1.0T/in² (15N/mm²) is also appropriate to the unreliability of cast iron in tension.

By treating the rail as continuous over a number of supports, the maximum bending moment is reduced to about seventy-five per cent of that obtained for a point load on a beam with two supports. Towards the end of the nineteenth century, rails were considered as being supported on a continuous elastic foundation. The theory of beams on elastic foundations was developed subsequent to Brunel's death, but his idea to adopt continuous support for the

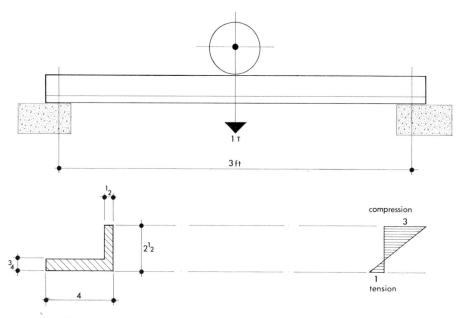

Fig 82 Plate rail with wheel load of 1 ton at midspan

rails lends itself to the application of the theory. It is possible to develop a relationship between wheel load and the resistance to deformation of the ground and the rail section. It can be shown (Timoshenko) that for rails of geometrically similar cross-section and a constant relationship between the material properties of the rail and ground, the maximum stress remains constant if the wheel load increases in the same proportion as the weight per unit length of the rail. Between 1838 and 1890, the weight of Brunel's bridge rail increased from 45lb per yard (22kg/m) to 120lb per yard (59kg/m) which reflected the increase in wheel load.

Column Tests

Brunel made extensive use of tests to verify his designs and the graph shown in Fig 25 illustrates some results of tests on timber columns which were carried out at Bristol in 1846 (I. Brunel). Over one hundred years previously, Leonard Euler (1707–83) derived an expression for the buckling load N_E of a hinged bar which is generally expressed in the following form:

$$N_E = \frac{\pi^2 E I}{L^2} \qquad \text{(equation 11)}$$

where E = modulus of elasticity
I = second moment of area
L = length between hinged ends

The corresponding compressive stress at the point of buckling for a cross sectional area A is

$$f_E = \frac{N_E}{A} = \frac{\pi^2 EI}{L^2 A}$$

For a square of side h the second moment of area I is $h^4/12$ and the critical stress can be written as

$$\frac{12\pi^2 E}{(\frac{L}{h})^2} \qquad \text{(equation 12)}$$

Thus it is possible to plot a relationship between the buckling stress f_E and the ratio L/h for a given value of the elastic modulus E for the material. Brunel obtained the value of E from bending tests. Equation 12 can be modified to take into account different end conditions and gives satisfactory results for valuves of f_E so long as this does not exceed the elastic limit of the material. Column design formulae used in modern timber engineering practice take into account initial lack of straightness and the creep of the material, that is, increasing deformation at constant load. Creep can be allowed for by a suitable reduction of the value of the elastic modulus.

Bending Moments

The variation in the value of the bending moment along beams with different support and loading conditions is shown in Fig 83. The maximum bending moment in a beam subjected to a total uniformly distributed load W over a span L between two supports is used as a basis for comparison. In all cases, the maximum load in any one span is W. It is apparent from the six cases shown in Fig 83 that a concentrated load acting at the end of a cantilever beam gives the maximum value of bending moment. The use of ballast to disperse concentrated wheel loads and reduce vibration, and the use of continuity over a number of supports are advantageous in terms of maximum values of bending moment. It must be remembered that the dispersion of the concentrated wheel loads is partially offset by the additional uniform loading due to the weight of the ballast. In the biography of Isambard Kingdom Brunel, his son wrote: 'Brunel was aware of the advantages of providing continuity in terms of bending moment and deflection values and carried out small scale tests on continuous beams to verify theoretical formulae.'

Analysis of Penadlake Viaduct

A preliminary structural analysis of the Penadlake Viaduct, of 1859 (see Fig 26), done by C. A. Mercer, revealed some interesting results. The main longitudinal beams consisted of two 24in (600mm) × 10in (250mm) beams connected together by means of bolts and joggles.

Joggles are small pieces of hardwood or cast iron of rectangular cross section, placed between two beams, and fitted carefully into notches cut across them. The beams are then bolted firmly

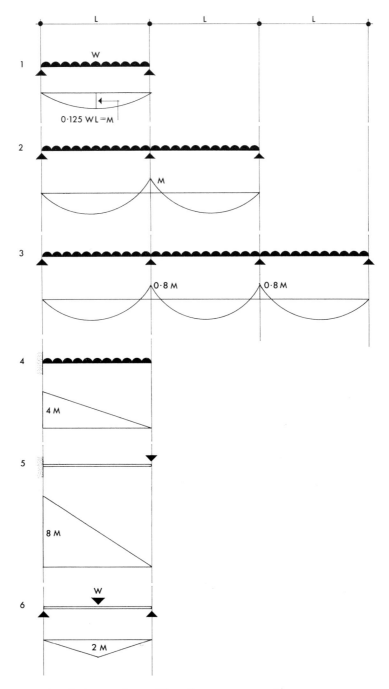

Fig 83 Comparison of bending moment values

together. The object is to stop the slipping which would occur between the two surfaces if the beams were merely laid one on the other and loaded, and so make two pieces of equal size act as one double the depth, and therefore four times the strength of a single piece. In 1841 Mr Brunel made experiments to satisfy himself of the strength gained by this method; and he afterwards perfected the arrangement by ensuring an exact fit in the notches by tightening up the joggles by wrought-iron wedges. (I. Brunel)

Equation 4 on page 201 shows that for a rectangular section, the load capacity in bending is proportional to its width times depth squared.

To obtain some experimental evidence of the efficiency of a bolt and joggle connection, Mercer carried out some small scale tests on beams of width b = 3.5in (87.5mm) and depth 1.5in (37.5mm) simply supported over a span of 4.875ft (1500mm) with a concentrated load W applied at midspan. The results are tabulated in Table 11 below and are related to the load required to induce a midspan deflection of span /300, that is 0.2in (5mm).

BEAM ARRANGEMENT	LOAD W lb(kN) required to induce a deflection of 0.2in (5mm)
1. Single beam width 3.5in (87.5mm) depth 1.5in (37.5mm).	89 (0.4)
2. Two beams as above placed one upon the other, no connection.	195 (0.87)
3. Two beams connected together with bolts and joggles.	337 (1.5)
4. Two beams glued together.	676 (3.01)

Table 11 The results of tests to demonstrate the increased stiffness obtained by a bolt and joggle connection

The results indicate that the bolt and joggle connection considerably increases the resistance to deformation. In theory, with no slip at the interface of the two beams, the resistance to deformation is proportional to the cube of the depth and it would be expected that the load W for case 4 is eight times that for case 1. In fact the experimental result was in the ratio 7.6 to 1, indicating a slip at the interface.

The main beams have an effective span of about 20ft (6.15m) between the raking supports and the load of a 60 ton (600kN) passenger train, when suitably dispersed, is equivalent to a uniformly distributed load of about 0.37T/ft (12 kN/m) or a total load W of 7.4T (74kN) on a span of 20ft (6.15m). From Fig 83 the simple beam moment M is given by

$$M = 0.125 \ W \ L$$

$$= 0.125 \times 7.4 \times 20$$

$$= 18.5 \text{ tons ft. } (56.9kNm)$$

Allowing for continuity this value can be reduced to 0.8M and the bending stress f is estimated from equation 5 on page 201.

$$f = \frac{4.8 \ M}{b \ h^2}$$

If the two beams are considered to act independently $h = 10$in (250mm) and

$$f = \frac{2.4 \times 18.5 \times 2,240 \times 12}{24 \times 10^2}$$

$$= 500\text{lb/in}^2 \text{ approx. } (3.45 \ \text{N/mm}^2)$$

Assuming full interaction between the two beams, this value reduces to 250lb/in² (1.73N/mm²) and for partial interaction, as would be the case with bolts and joggles, the result will be between these two values. The yellow pine used for Brunel's viaducts would have been capable of accepting a stress of 500lb/in² (3.45N/mm²) with a generous margin of safety, but moving trains induce much higher stresses and the influence of biodeterioration must also be considered. This simple analysis considers static loadings only.

The Suspension Chain

If the intensity of vertical loading w of a suspension bridge deck is constant along its length and the weight of the cable is small compared with this value, then the geometrical form taken up by the chain approximates to that of a parabola. Considering the equilibrium of the half chain in Fig 84, then by taking moments about support B

$$T_c = \frac{wL^2}{8h} = \frac{WL}{8h} \qquad \text{(equation 13)}$$

where W = total load on span L

At the support, the chain tension T_s is the resultant of $T_c = \frac{wL^2}{8h}$ and $V = \frac{wL}{2}$, thus

$$T_s = \frac{wL}{2} \left[1 + \frac{L^2}{16h^2} \right]^{1/2} \qquad \text{(equation 14)}$$

The above equations were used on pages 101 and 107 to estimate the chain tension in the Clifton Suspension Bridge.

It follows that by considering the arch as an inversion of the chain, the horizontal thrust in a parabolic arch of span L subjected to a uniform loading W is given by

$$H = \frac{WL}{8h.}$$

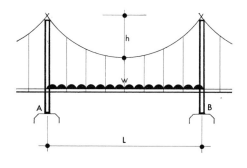

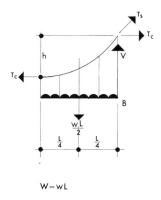

$$W = wL$$

Fig 84 The parabolic suspension chain

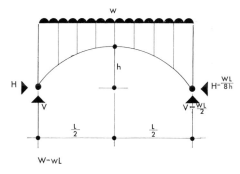

$$W = wL$$

Fig 85 The parabolic arch

This is indicated in Fig 85.

Choice of the correct ratio of L/h in suspension and arch bridge design is essential, and a ratio of 10 is commonly adopted for suspension bridges. A lower value of L/h is adopted for brick arches, in order to reduce the value of the arch thrust. The ratio L/h=5.28 adopted by Brunel for the Thames Bridge at Maidenhead was considered, at the time, to be dangerously high.

Roof Structure at Bristol Temple Meads

J. C. Bourne's description of the roof structure of Bristol Temple Meads Station as 'somewhat like the jib of a crane' (see page 135) is reasonably accurate and a simple mathematical model is shown in Fig 86. Considering a

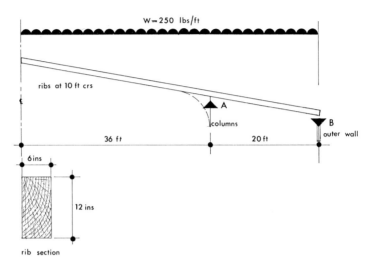

Fig 86 Bristol Temple Meads, roof structure

simplified loading condition of dead plus imposed (snow) load of say 25lb/ft² (1.25kN/m²) on the whole of the roof area, the load w per unit length is 250lb/ft. The rib is assumed to cantilever 36ft beyond the column line and thus the bending moment at A is given by

$$M_A = 250 \times 36 \times 18$$

$$= 162,000\text{lb ft}$$

The rib consists of a single 12in (300mm) by 6in (150mm) timber member with wrought-iron straps at the top and bottom. Considering the timber section only, the resulting stress is given by

$$f = \frac{6 \times 162,000 \times 12}{6 \times 12^2}$$

$$= 13,500\text{lb/in}^2 \ (93\text{kN/mm}^2)$$

Allowing for some reduction in the length of the cantilever due to the hammer-beam detail, this stress is still absurdly high thus suggesting the presence of substantial metal straps running along the top and bottom of the cantilever (Totterdill). For the loading shown in Fig 86, the column reaction A is given by

$$R_A = \frac{250 \times 56 \times 28}{20}$$

$$= 19,600\text{lb} \ (87.5\text{kN})$$

The uplift at the outer wall B is

$$R_B = 250 \times 56 - 19,600$$

$$= -5,600\text{lb} \ (-25\text{kN})$$

Bourne describes the 20ft (6.15m) short span or tail as being held down at the outer wall by a strong vertical tie passing some way down.

Tractive Effort

This is defined as the mean value of the force that the locomotive could exert on the rails when in full gear with maximum boiler pressure if the friction between coupled wheels and rails were great enough to transmit that force and if there were no friction anywhere else (Tuplin).

let $\quad$ P = maximum boiler pressure (lb/in²)
$\quad$ d = diameter of cylinders (in)
$\quad$ S = stroke of piston (in)
$\quad$ D = driving wheel diameter (in)
$\quad$ T = tractive effort (lb)

$$T \times \pi \times \frac{D}{2} = \frac{\pi}{4} \times d^2 \times P \times S$$

thus $\quad$ $T = \dfrac{0.5 \ P \ d^2 \ S}{D}$ lb

If we take the mean effective boiler pressure as 0.85 P the tractive effort reduces to

$$T = \frac{0.425 \ P \ d^2 \ S}{D} \qquad \text{(equation 15)}$$

The following data relates to Stephenson's 2–2–2 *North Star* (1837)

$$D = 7 \times 12 = 84\text{in}$$
$$P = 50\text{lb/in}^2$$
$$d = 16\text{in}$$
$$S = 16\text{in}$$

Thus tractive effort $\quad T = \dfrac{0.425 \times 50 \times 16^2 \times 16}{84}$

$$= 1{,}036\text{lb}$$

Relationship Between Piston Speed and Locomotive Speed

Using the following notation the relationship between piston speed and engine speed can be established:

$$V_T = \text{speed of locomotive mph}$$
$$D = \text{driving wheel diameter in feet}$$
$$S = \text{stroke of piston in feet}$$

Engine speed ft/min $\quad = V_T \times \dfrac{5{,}280}{60} = 88\,V_T$

Number of revolutions per minute R given by

$$R = \dfrac{88\,V_T}{\pi D} = 28.025\,\dfrac{V_T}{D}$$

The piston completes two strokes per revolution of wheel, thus piston speed V_p given by

$$V_p = 2SR\text{ft/min}$$
$$= 2 \times S \times 28.025\,\dfrac{V_T}{D}$$
$$= 56.05\,\dfrac{V_T S}{D} \qquad\qquad \text{(equation 16)}$$

Brunel originally proposed a piston speed of 280ft per minute, which for a stroke of 18in and a locomotive speed of 30mph gives

$$D = \frac{56.05 \ V_T \ S}{V_P} = \frac{56.05 \times 30 \times 1.5}{280}$$

$$= 9\text{ft}$$

Ship Stability

The stability of a ship is related to Archimedes' Principle which states that when a solid is immersed in a liquid, it experiences an upthrust equal to the weight. This principle is demonstrated quantitatively in Fig 87, in which an iron pontoon of overall dimensions 100ft × 20ft × 16ft deep weighing 450 tons is placed in fresh water of weight per unit volume 62.4 pounds per cubic foot. For flotation, the weight of the volume of water displaced must equal the weight of

The 450 ton pontoon displaces 8·08 ft of water

Fig 87
Fig 88

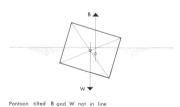

Pontoon tilted B and W not in line

the pontoon. Thus the depth x of water displaced is given by

$$x = \frac{450 \times 2{,}240}{20 \times 100 \times 62.4}$$

$$= 8.08\text{ft}$$

Thus the pontoon is in a stable floating position when the depth of water displaced (displacement) is just over one half the pontoon depth. The buoyancy of a body B is the upward force, represented by the weight of water displaced and is counteracted by the weight W of the ship acting downwards. For the level pontoon shown in Fig 87, these two forces are in line but if the pontoon is tilted the two forces are not in line (Fig 88). The relative positions of these two forces determine the stability of a ship. The rotational effect caused by the lines of action of the two forces not being coincident can produce a righting effect which tends to reduce the tilt, or the tilt can increase causing instability. In a similar manner it is necessary to consider the logitudinal stability of a ship and both longitudinal and transverse stability are influenced by the position of the cargo in relation to the line of action of the weight of the hull and superstructure.

BIBLIOGRAPHY

Banks, F. R. *The Penguin Guide to London* (Penguin, 1978)

Baxter, B. *Naval Architecture* (Teach Yourself Books, 1976)

Beckett, D. *Bridges* (Paul Hamlyn, 1969)

Body, Geoffrey, *Clifton Suspension Bridge, an illustrated history* (Moonraker Press, 1976)

Bourne, J. C. *The history and description of the Great Western Railway* (1846, reprinted David & Charles, 1970)

Brunel, I. *The Life of Isambard Kingdom Brunel Civil Engineer* (Longmans, 1870, reprinted David & Charles, 1971)

Clements, P. *Marc Isambard Brunel* (Longmans Green, 1970)

Clinker, C. R. *Paddington 1854-1979* – an official history of British Rail Western Region and Avon (Anglia Publications & Services, 1979)

Coleman, Terry. *The Railway Navvies* (Penguin, 1970)

Corlett, E. C. B. *The Iron Ship, the History and Significance of Brunel's Great Britain* (Moonraker Press, 1975)

Drysdale Dempsey, G. *Tubular Bridges circa 1850* (Kingsmead Reprints, 1970)

Drysdale Dempsey, G. *The Locomotive Engine circa 1850* (Kingsmead Reprints, 1970)

Emmerson, George S. *John Scott Russell* (John Murray, 1977)

Fuller, H. I. 'Steventon – story of a station', *Railway World* (April 1978)

Gilbert, K. R. *The Portsmouth Blockmaking Machinery* (HMSO, 1965)

Gimpel, J. *The Medieval Machine* (Tutuna Publications, 1979)

Gladwyn, Cynthia. 'The Isambard Brunels', paper read at a joint meeting of the Institution of Civil Engineers and the Societe des Ingenieurs Civils de France, *Proceedings of the Institution of Civil Engineers* (1970)

Hadfield, C. & Skempton, A. W. *William Jessop, Engineer* (David & Charles, 1979)

Hartwell, R. M. (Editor). *The Causes of the Industrial Revolution in England*, (Methuen, 1967)

Law, R. J. *The Steam Engine* (HMSO, 1977)

Lee, Charles, E. *The East London Line and The Thames Tunnel* (London Transport Publication, 1976)

Macdermot, E. T. *History of the Great Western Railway*, Volumes 1 & 2 (Ian Allan, 1964)

Marshall, J. *The Guinness Book of Rail Facts & Feats* (Guinness Superlatives, 1975)

Mercer, C. A. 'The Penadlake viaduct built by I. K. Brunel', final year undergraduate project, Department of Civil Engineering, University of Surrey

Meriam, J. L. *Statics* (John Wiley & Sons, 1966)

Morgan, Bryan. *Railways: Civil Engineering* (Arrow Books, 1973)

Pearce, David & Binney, Marcus (Editor). *Off the Rails: Saving Railway Architecture*, (SAVE Britain's Heritage, 1977)

Pugsley, Sir Alfred. *The Theory of Suspension Bridges* (Edward Arnold, 1968)

Pugsley, Sir Alfred (Editor). *The works of Isambard Kingdom Brunel* (Institution of Civil Engineers and University of Bristol, 1976)

Rawson, K. J. & Tupper, E. C. *Basic Ship Theory*, Volume 1 (Longman, 1976)

Reed, P. J. T. *The Locomotives of the Great Western Railway* Part Two, Broad Gauge (Railway Correspondence and Travel Society, 1952)

Rolt, L. T. C. *Isambard Kingdom Brunel* (Longmans Green, 1957; Pelican, 1970)

Russell, W. H. *The Atlantic Telegraph* (1865, reprinted David & Charles)

Singer, Charles, *et al. A History of Technology*, Volume 4, (Oxford University Press, 1975)

Slinn, J. N. *Great Western Way* (Historical Model Railway Society, 1978)

Smiles, Samuel. *Lives of the Engineers*, Volume 3 (David & Charles reprint, 1969)

Stevens, T. & Barnes, G. W. *The history of the Clifton Suspension Bridge* (The Clifton Suspension Bridge Trust)

Sylvester Charles. *Report on Rail-Roads & Locomotive Engines* (1825, reprinted E. & W. Books, 1970)

Thomson, David. *England in the 19th century* (Penguin, 1977)

Timoshenko, S. P. *History of strength of materials* (McGraw-Hill, 1953)

Totterdill, J. W. '. . . a peculiar form of construction', *Journal of the Bristol and Somerset Society of Architecture*, Vol 5

Tuplin, W. A. *British Steam Since 1900* (David & Charles, 1969)

Vaughan, A. *A Pictorial Record of Great Western Architecture* (Oxford Publishing, 1977)

Westcott, G. F. *The British Railway Locomotive 1803–1853* (HMSO, 1977)

Whitley, H. S. B. 'Timber viaducts in South Devon and Cornwall, GWR' *The Railway Engineer* (October 1931)

Wishaw's Railways of Great Britain and Ireland (1842, reprinted David & Charles, 1969)

INDEX